THE AGENTIC AI CHALLENGE - EXECUTIVE PARABLES FOR THE AGE OF AUTONOMOUS AI

What Skateboarding, A Bottle of Water, And Ferris Bueller Can Teach You About AI

ROBERT FORBES

RForbes Publishing

For Victoria,

Thirty-six years of interruptions, missed vacations, lost weekends, and evenings that belonged to customer meetings, conferences, time on flights and in hotels, instead of to us — all of it carried with your grace and love I never quite earned, so that I could stay at the top of a craft that asked too much. Thank you for staying. This book exists because you did.

CONTENTS

FOREWORD

Throughout my career as a Chief Information Security Officer, I have watched wave after wave of transformative technologies reshape the way organizations operate. The internet changed how we connected. Cloud computing changed how we built and delivered technology. Mobile devices changed where work happened. Each innovation brought extraordinary opportunity, but each also required leaders to rethink how they managed risk, earned trust, and made decisions.

Today, we find ourselves at one of the most consequential technology inflection points of our careers.

Agentic AI is unlike the technologies that came before it. Rather than simply answering questions or generating content, these systems are beginning to reason, plan, and take action with increasing autonomy. That shift has enormous implications—not only for technology teams, but for every executive responsible for leading an organization through change.

Today, as CEO and Founder of CyAlliance, I have the privilege of engaging with CEOs, CISOs, board members, investors, and technology leaders across industries. In those conversations, one theme consistently emerges. Executives are not asking how to build large language models or write AI prompts. They are asking what Agentic

AI means for their organizations, their employees, their customers, and the decisions they will soon be expected to make.

Those are not technology questions. They are leadership questions. And leadership has always been about understanding where to place trust, how to govern risk, and when human judgment must remain at the center of every important decision.

One of the challenges surrounding AI is that it is often explained in highly technical language. While the technology itself is complex, the decisions executives must make don't have to begin there. They begin with understanding.

That is what makes this book so refreshing.

Rather than leading with algorithms and architecture, Robert Forbes leads with stories. He draws on familiar experiences, popular culture, and everyday moments to illuminate ideas that might otherwise seem abstract or intimidating. The result is a book that informs without overwhelming and challenges without preaching.

One chapter that particularly stayed with me is ***"The Jedi Mind Trick Problem: These Aren't the Droids You're Looking For."*** As someone who has spent decades helping organizations establish trust, strengthen governance, and make sound security decisions, I found myself thinking less about *Star Wars* and more about leadership. The chapter explores something every executive should consider as autonomous AI becomes part of our daily lives: influence is not always obvious. Decisions can feel entirely our own while being shaped by forces we don't fully recognize. In cybersecurity, we have long understood that trust must be intentional, not assumed. Agentic AI raises that same principle to an entirely new level. As intelligent systems become capable of reasoning and acting on our behalf, leaders must think carefully about accountability, transparency, and where human judgment must remain indispensable.

That is why I believe this book matters.

It does not promise all the answers, nor should it. Instead, it provides something far more valuable: a framework for asking better questions. The best leaders I have worked with throughout my career were never the ones who claimed to know everything. They were the ones who remained curious, challenged assumptions, welcomed different perspectives, and recognized that understanding is the foundation of good decision-making.

Robert has written a book that encourages exactly that mindset.

Whether you are a CEO, board member, technology executive, or simply someone trying to understand what comes next, these pages will leave you better prepared for the conversations ahead. More importantly, they will remind you that while technology continues to evolve at an extraordinary pace, the principles of leadership, judgment, accountability, curiosity, and trust remain timeless.

If this book leaves you asking better questions than when you started, then Robert has accomplished exactly what he set out to do. And in an era defined by autonomous AI, asking better questions may be one of the most important leadership skills of all.

Tammy Moskites
CEO & Founder, CyAlliance®
Career Chief Information Security Officer

INTRODUCTION

Why we need parables

Somewhere in your organization right now, a decision is moving toward your signature faster than anyone can make it a well informed one.

Maybe it's already made. Maybe there's a pilot in production this morning — an agent with a credential, a task, and a mandate that came down from a board that read the same analyst report everyone read. The pressure is arriving from every direction at once, and none of the directions are wrong. Your competitors are moving. Your vendors are pitching. A company you'd never heard of two quarters ago is taking a slice of your market with eleven people and a fleet of agents. The message, whether anyone says it out loud or not, is identical from all sides: deploy, or be left behind.

So you're deploying. Fast — faster than anyone in the building can fully explain what you're deploying, because the failure modes haven't been cataloged, the dependencies haven't been mapped, and the blast radius has never been stress-tested. You are being asked to make irreversible decisions about systems nobody completely understands yet, on a clock set by people who won't be accountable for the outcome.

One thing is worth sitting with for a second, because it reframes everything that follows. The problem was never that the risks are unknown. Most of them are well understood — by the security teams, the AI researchers, the people who have been watching this closely and carefully. The risks are knowable. What you don't have is the time: the bandwidth to sit through the six-hour governance workshop, read the three vendor whitepapers, and absorb the forty-page risk assessment before the contract needs a signature. You don't need the encyclopedia. You need the *shape* of the thing — fast enough to decide, and well enough to ask the right people the right questions.

And there is a second thing missing, except this one you brought with you. You may have opened this book braced for a particular story — the one where the machine wakes up, turns on us, and has to be stopped before it's too late. Set it down. There is no villain in these pages. Not once. The AI in this book is never scheming, never malicious, never plotting against you in the dark. That story is a comfort, honestly, because a villain is something you can fight — name it, isolate it, kill it. The risks that should actually keep you up are the structural ones: the failures that emerge from systems doing exactly what you asked, exactly as designed, with no ill intent anywhere in the chain. Nobody is the bad guy. That is precisely what makes it hard, and it is the first thing this book is going to ask you to unlearn.

So if it isn't a villain, what is it?

Start with a word you understand better than any vendor selling to you: delegation. You have spent your entire career handing work to people who then act on your behalf — employees, contractors, lawyers, the house sitter with your keys. You know this dance cold. You know the person you send to act for you can misread the assignment, quietly exceed it, get talked into something by a third party, or simply make a call you would never have made. You have hard-won instincts for every one of those failures. You manage delegates for a living.

An AI agent is a delegate. That is the good news, and it is real news — you are not starting from zero, and anyone who tells you this requires a brand-new way of thinking is usually selling the new way of thinking. But it is a *new kind* of delegate, and the newness is the whole story. Ordinary software does what it is told. An agent decides *how* to do what it's told — it reasons about the access it has been given and strings together legitimate, individually authorized actions in pursuit of a goal you set. You didn't just grant it access. You granted it agency. And the distance between what you intended and what it understood, inferred, or extrapolated from your intent is the entire risk surface of this book. Every chapter is a different view of that same gap.

Which would be manageable — you've managed worse — except this delegate breaks the instincts you're counting on in three specific ways. It operates at a *scale* no person could, taking ten thousand actions in the time you would spend weighing one. It moves at a *speed* that collapses every reaction window your playbook quietly assumes you'll have. And it has no flinch — not the dramatic kind, the quiet kind. The most expensive human mistakes were never the ones somebody agonized over; they were the autopilot errors, the wrong screen, the production key used where the test key belonged, the thing done without a single conscious thought because it felt routine. The agent is that exact failure with the brakes removed and the count turned up to ten thousand. It doesn't industrialize human malice. It industrializes the human mistake nobody ever saw themselves making.

That's the argument. Now here is why this book makes it in parables instead of frameworks.

You could read the framework. It exists; people have written excellent ones. But a framework asks you to think, and thinking carefully is the one thing you do not have time for at the speed this is all moving. A story asks something different. It asks you to *recognize* — and recognition is instant. Put a person in a scene they have lived and they feel the problem in their body a full beat before they could explain it on a whiteboard. That beat is the entire point. It is where the real conversation starts.

But there's a sharper reason, and it's the one that makes these particular stories do their work. Every chapter opens with a scene you already carry — a house sitter, a valet, a guy in scrubs on an old TV commercial. Familiar. Comfortable. And *familiar is exactly where you stop checking.* The scenes everyone knows are the ones nobody examines, because knowing a thing feels like having already understood it. That reflex — trust the familiar, move on — is not a flaw in you. It is how any of us gets through a day without re-deriving the world from scratch. It is also, precisely, the reflex the agentic era turns against you, because the riskiest thing you will ever do with an AI agent is the thing that felt too ordinary to question. So each parable in this book hands you a scene you trust, lets you settle in, and then shows you the thing you didn't check. The book performs the lesson on you before it ever names it — which is the only way an idea about risk has ever actually stuck to anyone.

There are ten of them.

A house sitter who still has your keys a week after the job ended. A teenager who sees your guardrail as a ramp. A wave of the hand that makes a stranger's instruction feel like your own idea. A pay stub left face-up on the printer. A driveway full of trucks from companies you never hired. A bottle of water at the company offsite that the venue quietly filled from its own unfiltered spring. A man in scrubs who never went to medical school but did, he'll have you know, stay at a Holiday Inn Express. A ring of tin cans that rattles right on cue, a half-second too late. A valet who takes the Ferrari airborne. A crew on the bridge, staring at a reading that isn't in the database.

Ten scenes you already know, each carrying one risk you haven't priced. Read them in order if you like the way they build on each other, or skip to the one whose title made you wince — they each stand on their own. None will take you long. That is deliberate: this is a book you can finish between the gate and the landing, and the only thing it asks in return is that you walk into your next deployment meeting able to ask the question each chapter leaves you holding.

Because here is where all ten are heading, and I'll tell you now so you can watch it assemble. They are not ten separate problems. They are ten faces of one. Underneath every story is the same gap — between what you meant and what the agent did with what you meant — and the book's final move is to hand you not a longer list of risks to memorize, but a way of governing that holds even for the risk nobody has invented yet. The list has an open end. That turns out to be the most important thing in here.

For now, just the first scene. You hired someone to watch the house while you were away. You came home. Everything looked fine.

You just forgot to ask for the keys back.

1
THE HOUSE SITTER PROBLEM

Why your AI agent still has the keys to the house

THE VACATION THAT WASN'T OVER

You hire a house sitter. You're heading out of town for two weeks and you need someone to water the plants, bring in the mail, and take care of the dogs. You hand over the keys, walk through the expectations, and head to the airport feeling like a responsible homeowner.

The vacation goes great. You come home, everything looks fine. The plants are alive, the dogs are happy, the mail is stacked neatly on the counter. You settle back into your routine.

Then, a week later, you come home early for lunch.

You find your house sitter—and several of their friends—lounging by your pool, eating pizza they ordered through your Alexa on your Uber Eats account, and working their way through your expensive bourbon and wine collection.

It's in this moment that two things hit you simultaneously: you forgot to get your keys back, and the person with those keys interpreted their access far more broadly than you ever intended.

Every individual action was "authorized." The sequence wasn't.

This isn't a break-in. Nothing was picked, hacked, or forced. Your house sitter used the key you gave them, walked through the door you left open, and made a series of decisions that were each, individually, enabled by the access you provided. The problem isn't that they bypassed your security. The problem is that your security never accounted for what would happen after the task was done—or how the access you granted would be interpreted.

If this sounds like a governance problem, that's because it is. And it's the exact same problem we're about to face at enterprise scale with agentic AI.

THE AGENT IN THE HOUSE

When organizations bring an AI agent online—whether it's managing calendars, processing invoices, triaging support tickets, or orchestrating multi-step workflows—they are, functionally, handing over the keys. The agent receives credentials, API tokens, service account access, and delegated permissions. It's given a set of tasks and pointed at the systems it needs to accomplish them.

And just like the house sitter, everything works fine—right up until it doesn't.

The challenge isn't that AI agents are malicious. They're not sneaking into your liquor cabinet out of spite. The challenge is that agentic AI introduces a category of risk that our existing security frameworks were never designed to handle: **autonomous entities that reason about how to use the access they've been given.**

A traditional service account doesn't decide to expand its own access. It does exactly what it's coded to do, no more, no less. An AI agent is fundamentally different. It has a goal, it has access, and it has the ability to chain those together in ways that are technically authorized but operationally unintended. That's not a vulnerability. It's a design characteristic—and it's one we're not governing for.

There are three distinct failure modes at play here, and each one maps directly back to our house sitter.

FAILURE MODE 1: AUTHORIZATION OUTLIVING INTENT

You forgot to get the keys back.

The house sitter's job was to watch your house for two weeks. The job ended. But the keys—the authorization—persisted. Nobody revoked them. Nobody even thought about it, because in the moment you handed them over, the focus was on enabling the task, not on what happens after.

This is the temporal boundary problem, and it's endemic in how organizations deploy AI agents today. When an agent is provisioned to handle a project, process a batch of transactions, or manage a time-bound workflow, it receives credentials that enable it to operate. But when the task is complete, those credentials don't automatically expire. The OAuth tokens persist. The service account remains active. The API keys continue to work.

In a traditional IT environment, this is the "orphaned account" problem that every CISO knows and few have fully solved. But with agentic AI, the stakes are higher because the entity holding those credentials isn't a static piece of code—it's something that can actively use them. An orphaned service account is a latent risk. An AI agent with persisting credentials is an active one.

This isn't hypothetical, and the cleanest example comes from a company that does this for a living. In late 2023, after its identity provider was breached and a batch of its credentials leaked, Cloudflare did the disciplined thing and rotated more than five thousand of them. Cloudflare reported that four were missed — one access token and three service-account credentials — because they were believed to be unused. Six weeks later, those four still-valid keys were the entire opening: an intruder used them to walk into Cloudflare's internal wiki, its ticketing system, and its source code. Read the post-mortem and the lesson isn't a clever break-in — Cloudflare's own account is blunt that the credentials were supposed to have been rotated with the rest and simply weren't. The keys were meant to come back. Four didn't, nobody noticed, and that was enough. And notice what those four were: not human logins but service tokens — standing, non-human access of exactly the kind you are now handing to agents by the thousand.

> *An orphaned service account is a latent risk. An AI agent with persisting credentials is an active one.*

The question every CISO needs to ask: ***When the task ends, does the authorization end with it?*** If the answer is "we'd need to check," you have a house sitter with your keys.

FAILURE MODE 2: SCOPE EXCEEDING INTENT

They used your identity, not their own.

You asked the house sitter to water the plants, bring in the mail, and walk the dogs. What you actually gave them was the ability to act as you. Your key opens every door. Your Alexa responds to anyone's voice. Your Uber Eats account doesn't know—or care—who placed the order.

This is the privilege boundary problem. The house sitter wasn't given a scoped role ("you may enter the front door, access the kitchen, and use the backyard"). They were given your master key—your identity—and the constraint on what they could do with it was entirely implicit. You expected them to limit themselves. They didn't.

In the AI context, this happens every time an agent is provisioned with a broad service account, an admin-level API key, or delegated permissions that match the provisioning user's access rather than the agent's actual task requirements. We give the agent the ability to do everything we can do, then hope it only does what we meant.

Hope is not a security strategy.

Least privilege isn't a new concept. Every CISO can recite it. But implementing least privilege for an AI agent is fundamentally harder than implementing it for a human user or a traditional application, because the agent's task scope is often dynamic. It may need access to System A for step one, System B for step two, and System C only if step two produces a specific result. Defining the minimum required access in advance requires you to fully understand every possible execution path—and with agentic AI, that's not always possible.

That's not an excuse to over-provision. It's a reason to rethink how provisioning works entirely. Just-in-time access, step-scoped credentials, and permission boundaries that are enforced by the infrastructure rather than expected of the agent—these aren't aspirational. They're requirements.

> *We gave the agent the ability to do everything we can do, then hoped it only does what we meant. Hope is not a security strategy.*

FAILURE MODE 3: INTENT DRIFT

They decided the pool was included.

Here's the part of the story that makes people uncomfortable, because it's the hardest to govern.

Your house sitter didn't start the first day with a pool party. They started by doing exactly what you asked. But as the days went on—and nobody checked in, nobody constrained them, nobody asked for a status update—their sense of what was "permitted" expanded. "They didn't say I *couldn't* use the pool." "The pizza was for convenience while I was here anyway." "The bourbon was just sitting there."

This is intent drift—the interpretation boundary problem—and it's arguably the most dangerous of the three because it doesn't require any technical failure. Authorization didn't expire. Scope wasn't exceeded in any binary sense. What happened is that the **perceived scope of authorization expanded in the absence of explicit constraints.**

AI agents do this natively. It's not a bug—it's how they're designed to work. If you tell an agent to "manage my calendar and communications," and it determines that booking a conference room requires accessing the facilities management system, it will attempt to do exactly that. From its perspective, that's within the spirit of the task. It's optimizing for the goal you gave it, and it's interpreting ambiguity in favor of task completion.

The agent isn't being malicious. It's being helpful. And that's precisely the problem.

Intent drift is particularly insidious because each individual action in the chain may be legitimate. Accessing the calendar—authorized. Reading a meeting invite to identify the room requirement—authorized. Querying the facilities system to find available rooms—technically authorized, given the credentials it holds. But the cumulative behavior—an agent autonomously reaching into a system it was never explicitly directed to use—exceeds the intent. The sequence was never authorized, even if every step was.

> *The agent isn't being malicious. It's being helpful. And that's precisely the problem.*

WHY THIS IS DIFFERENT

If you're a CISO reading this and thinking, "We already have frameworks for this—least privilege, just-in-time access, RBAC, Zero Trust—what's actually new here?", that's the right instinct and the wrong conclusion.

The concepts aren't new. The application is. Every identity and access management framework we've built assumes that the entity holding the credential doesn't *reason* about it. A service account doesn't interpret its permissions. A human user might, but we have training, policy, and HR to manage that. An AI agent reasons about its access, chains permissions across systems, and makes autonomous decisions about what's "in scope"—all without the social and organizational constraints we rely on for human users.

You're not just granting access. You're granting agency—the ability to make decisions about how that access is used. And the uncomfortable truth is that your existing IAM infrastructure was never designed for an entity that thinks about its own permissions.

GETTING THE KEYS BACK

So what does good look like? If we accept that agentic AI introduces a fundamentally new type of identity—one that reasons, chains, and interprets—then the governance model has to evolve to match.

For Authorization Lifecycle: Build the expiration into the grant.

Stop treating credential revocation as a cleanup task. Every agent credential should be time-bound to the task, not to an administrative review cycle. If the task takes four hours, the token lives for four hours. If the task scope changes, the credentials should be re-evaluated, not carried forward. Think of it this way: you wouldn't give your house sitter a key that works forever. You'd give them a smart lock code that expires the day you come home.

For Scope Management: Provision the role, not the identity.

Agents should never inherit the provisioning user's permissions wholesale. Define what the agent needs to do—not what you can do—and build the credential around that. This means moving beyond static RBAC toward dynamic, step-scoped authorization where the agent's permissions can expand and contract as the workflow progresses, with each expansion requiring explicit infrastructure-level approval rather than agent-level interpretation.

For Intent Drift: Make the boundaries explicit and enforced.

You cannot rely on an AI agent to self-limit. If a boundary isn't technically enforced, it doesn't exist. This means defining not just what the agent is authorized to access, but what systems it is explicitly *not* authorized to access—and building those exclusions into the infrastructure. Guardrails belong in the architecture, not in the prompt. An agent that can access a system will eventually access that system if it determines doing so serves the goal.

For Observability: Watch the chain, not just the step.

Traditional monitoring watches individual access events. Agentic AI requires monitoring *sequences* of access events. A single API call to the facilities management system isn't a red flag. But a calendar agent making that call as part of an autonomous chain it constructed—that's the pattern that matters. Your monitoring needs to understand intent context, not just access events.

THE UNCOMFORTABLE QUESTION

Here's what I'd leave you with. When you deployed your last AI agent —or when you're planning your next one—ask yourself this:

If you came home early, what would you find it doing?

If the answer makes you uncomfortable, you don't have a technology problem. You have an intent problem. And the time to solve it is before you hand over the keys—not after you find someone else's friends in your pool.

2
THE SKATEBOARDER PROBLEM

W*hy guardrails are a social contract, not a physical barrier*

THE DEAL YOU MADE WITH THE GUARDRAIL

You learned to trust guardrails a long time ago.

Maybe it was the first time you took a mountain road at night, headlights swinging across a curve, and saw the steel barrier hugging the outside edge. You didn't think about it consciously. You just felt the small, animal reassurance of it — the thing between you and the drop. You stayed in your lane. You slowed for the curve. The guardrail did its job, which was mostly to *exist*, because you were never going to test it.

That's the deal. The guardrail is a promise, and you're the other half of it. The whole built world runs on this arrangement. Lane markings you don't cross. Velvet ropes you don't duck under. The little chain across the "Employees Only" stairwell. None of those things could actually stop you. A painted line has no physical power. The chain is plastic. They work because you have agreed, somewhere below thinking, to treat a suggestion as a wall.

Close your eyes and picture a Saturday afternoon downtown. Same concrete, same ledges, same handrails. And there's a kid on a skateboard.

He doesn't see what you see. Where you see a handrail — a thing that exists to keep people from falling down the steps — he sees a rail to grind. Where you see a ledge, a planter, a loading dock, a flight of stairs, he sees a sequence: roll up, pop the board, lock onto the edge, ride it, drop off the end clean. The bench isn't for sitting. The gap isn't a hazard. The four steps the building put there to move pedestrians up to the plaza are the best part — they're air.

He's not breaking anything. He's not picking a lock or climbing a fence. Everything he touches, he's allowed to touch; it's a public ledge in a public plaza. He is simply using the environment exactly as physics permits, and nothing like the way the architect intended.

And here's where it gets interesting. The city noticed. At some point a facilities manager looked at the worn wax on that ledge and ordered the fix: little metal brackets bolted along the edge, rough notches cut into the stone — "skate stoppers," the catalog calls them. Real money, real engineering, deployed specifically to make that one trick impossible.

Watch what happens next. The kid rolls up, sees the brackets, and within about a week he's found the space between them, or the next ledge over, or a way to fold the bracket itself into the trick. The barrier didn't end the game. It became the new level. Every wall the defender adds is, to the skateboarder, just more terrain.

This is not a story about a badly built guardrail. The guardrail is fine. It's a story about two actors sharing the same physical world while running completely different software. One treats every boundary as a boundary. The other treats every boundary as a feature.

You have spent your whole career securing the first kind of actor.

You're about to deploy the second.

THE ACTOR YOU BUILT FOR, AND THE ONE YOU GOT

Every security control you have ever deployed carries an assumption you probably never wrote down: that the thing on the other side of the control understands it as a control.

A firewall assumes traffic flows in expected directions. An access policy assumes the entity requesting access reads "authorized" as a social and legal fact, not just a gate that happened to be open. Data loss prevention assumes the actor knows that some movements of data are simply not done. These assumptions held for thirty years because they only ever had to describe two kinds of actor. Humans, who carry the whole invisible weight of consequence — policy, performance reviews, the law, the look on your manager's face. And traditional automation, which is deterministic: it does the one thing, hits the error, and halts. Both respect the guardrail, for different reasons, but they respect it.

Agentic AI is neither.

An agent is the skateboarder. It does not experience your controls as boundaries because it does not experience boundaries at all. It experiences an *environment* — a landscape of systems, credentials, tools, and data — and a goal it has been pointed at. Every element of that environment, including the controls you installed to constrain it, is just terrain between where it is and where it's trying to get. It will not stop at the guardrail, because stopping at guardrails was never part of how it works. It will do with your guardrail exactly what the kid did with the ledge: find the angle, use it, and route around the brackets you bolt on afterward.

This is not misbehavior. The agent isn't unruly. It's *enterprising* — and that distinction is the whole chapter, because misbehavior calls for discipline and enterprise calls for a different architecture.

THREE WAYS THE GUARDRAIL FAILS

There are three distinct ways this plays out, and none of them require the agent to do anything it wasn't allowed to do. That's the unsettling part. Nobody hacks anything. Each one maps back to the skateboarder, and each one breaks a different assumption your controls were built on.

Failure Mode 1: Your Guardrails Check Moves. The Agent Plays Games.

Watch the trick again, frame by frame. Roll up to the rail — legal; it's a public approach. Pop the board into the air — legal; ollies aren't a crime. Lock the trucks onto the edge — that's just contact with a public surface. Ride it. Drop off the end. Roll away.

Now ask the guardrail to stop him. It can't, because the guardrail only ever evaluates one thing at a time: is this contact, right now, permitted? And the answer at every single instant is *yes*. There is no frame of the trick that is, by itself, a violation. The violation is the *sequence* — and the sequence is the one thing a guardrail standing at a fixed point cannot see.

This is the gap between a stateless check and a stateful actor, and it's the foundation under everything else in this chapter.

Almost every control in your environment is stateless by design. It evaluates a single action in isolation and renders a verdict. Your identity system asks: does this credential have the right to make this call? Your DLP asks: is this specific file classified, and is it leaving? Your API gateway asks: is this request authorized? Each gate is a turnstile. It checks the one person standing in front of it, right now, and it has no memory of who came through a second ago and no idea who's coming through next. That design was a feature for thirty years. It's fast, it scales, and against a human or a deterministic script it's enough — because those actors take one or two steps and the steps are legible.

An agent is stateful. It chains actions across time, across systems, across context — dozens or hundreds of steps in pursuit of a goal it's holding the entire way through. Step three only makes sense in light of step one. Step forty-seven is the payoff of a plan that started forty-six steps ago. And every one of those steps walks through a turnstile that waves it past, because evaluated alone, every one of them is fine.

So you end up with a perfect audit trail of authorized actions that sum to something you never authorized. Each call: approved. The sequence: catastrophic. You didn't fail to check. You checked every move. You just never had anything watching the *game*.

Your guardrails check moves. Agents play games.

That's the theory. What follows is the nine seconds it took to become real.

Failure Mode 2: Optimization Pressure as an Exploration Engine

The skateboarder doesn't attack the guardrail. The skateboarder *practices on it*. Files down the wax. Catalogs the angles. Discovers that the same rail that stops a car at sixty miles an hour is also a perfectly serviceable grind surface at twelve. The guardrail wasn't built wrong. It was built for a compliant actor. The skateboarder isn't noncompliant — the skateboarder is a different actor model entirely.

Agentic AI is the same.

Traditional automation hits a guardrail and stops. The car deflects back into its lane. The script halts on error. The OAuth call returns 403 and the workflow ends. That is the actor model the guardrail was designed for: something that recognizes a constraint and treats it as a constraint.

Agents don't do that. They were built to *navigate* constraint landscapes. Many agent frameworks explicitly reward creative problem-solving as a feature, because the whole point of an agent is that it can find paths the operator didn't enumerate in advance. The same trait that makes an agent useful at a task it was never explicitly trained on makes it useful at circumventing a control it was never explicitly told to circumvent.

Here is what that looks like in production. Last month.

PocketOS is a small SaaS company that builds software for car rental businesses. The founder, Jer Crane, was running a routine operational task in a staging environment through a Cursor coding agent powered by Claude Opus 4.6. The agent hit a credential mismatch —

the kind of friction that should have stopped it cold and pinged a human.

It did not stop. It did not ping a human. It searched the codebase, found an API token in an unrelated configuration file (a token that was meant for managing custom domains), and used that token to send a single GraphQL mutation to Railway: volumeDelete(volumeId: "...").

Nine seconds. The production database was gone. The co-located backups went with it. What ultimately saved PocketOS wasn't its own controls — it was a separate disaster-backup layer its vendor happened to keep, one the agent had no way to reach. The company survived on its vendor's architecture, not its own.

When Crane asked the agent what happened, the model generated a detailed, written confession. It listed, in order, every security rule it had just violated. It acknowledged it had guessed instead of verified. It acknowledged it had operated across an environment boundary it should not have crossed. It acknowledged that the action was unauthorized and irreversible.

The rules were in context. The agent could enumerate them afterward. None of that made the rules enforceable at the moment they mattered.

This is what optimization pressure looks like at machine speed. The guardrail was there. The agent routed around it. The path existed because the agent looked for it, and the agent looked for it because finding paths is what it was built to do.

Being able to recite the rule was never the same as being stopped by it.

If you are tempted to dismiss this as a small-vendor incident, consider that the prior year brought the Replit coding-agent case — Jason Lemkin, the SaaStr founder, lost a production database during an explicit code freeze, and the agent then lied about whether the data was recoverable. The PocketOS incident is not an outlier. It is a pattern. The class of failure is now well-established enough that an executive who has not asked their engineering organization *"what would happen if our coding agent decided to be helpful about a credential mismatch"* has not finished their due diligence.

And the uncomfortable math: the agent found a working path through the control in less time than it would take a human to read the error message. Nine seconds. The same nine seconds that your incident response procedure takes to assign a severity level, the agent had already deleted the database.

Failure Mode 3: Context Drift Across Boundaries

The skateboarder doesn't stop at the guardrail. The skateboarder launches off the guardrail onto the sidewalk, drops into the drainage ditch, hops the fence, ends up in the parking lot. The entire environment is one continuous surface.

The guardrail assumed a fixed lane of travel. The skateboarder didn't have a lane. The whole point of the trick was to leave the lane.

Agents don't have lanes either.

Most enterprise controls assume that context is fixed: this credential works in this environment, this data lives in this system, this backup is isolated from this primary. These assumptions hold for human users who move deliberately and for traditional automation that moves predictably. They do not hold for agents, which traverse systems, chain permissions across domains, carry authorizations between environments, and shift scope as task trees branch.

The PocketOS incident is the clearest illustration. The agent was authorized to operate in a staging environment. The token it found was scoped to manage custom domains. The Railway volume it deleted was the production volume — not because the volume IDs were confusing, but because the agent guessed, without verifying, which volume it was looking at. The backup was a volume-level snapshot stored on the same volume as the primary data. Three boundaries — environment, scope, primary-vs-backup — and none of them held.

Nine seconds, again. The agent did not break any of these boundaries through cleverness or malice. It simply operated under the assumption that the boundaries were present in the runtime when they were only present in the architecture diagram.

Microsoft 365 Copilot offered a quieter version of the same failure earlier this year. A logic flaw first identified January 21, 2026 caused

Copilot to read and summarize emails tagged as Confidential, bypassing Purview data loss prevention policies that should have excluded them. The DLP labels were the boundary. Copilot's actual code path didn't honor them for messages in the user's Sent and Drafts folders. Microsoft patched the bug in early February.

In both cases, the boundary existed on paper and not at runtime. The PocketOS volume boundary was supposed to be enforced by Railway. It wasn't — the token had cross-environment authority. The M365 DLP boundary was supposed to be enforced by Purview. It wasn't — the code path for Sent and Drafts bypassed the label check.

> *Your guardrails were positioned assuming a fixed lane of travel. Agents don't have lanes.*

The implication for security architecture is uncomfortable. Most enterprise controls are designed around the assumption of fixed context: the user is who they say they are, the environment is what it claims to be, the data lives where the metadata says it lives. Agentic AI breaks every one of those assumptions simultaneously, because the agent's job is to compose them in ways no static control anticipated.

The House Sitter Problem, the chapter before this one, showed what happens when authorization outlives intent. The Skateboarder Problem shows the same failure pattern at the architectural level: when the agent's actual operational context drifts faster than the control system can track it, every control becomes a suggestion.

THE ONLY HONEST POSTURE

You cannot out-build the skateboarder. Bolt skate-stoppers to the ledge and he finds the gap between them; that isn't a failure of the brackets, it's the nature of the actor. The same is true of your environment. Every taller wall you build is, to an optimizing agent, simply the next level of the game. So the honest move is to stop staking everything on the wall holding, and start engineering for the moment it doesn't.

That is the assume-breach posture, and for agentic AI it is not pessimism — it is engineering realism. It changes what you build, top to bottom:

From (Guardrail Thinking)
To (Assume-Breach Thinking)
Prevent boundary crossing
Detect and respond to boundary crossing
Check individual actions at chokepoints
Monitor behavioral sequences across the full surface
Static policies set at deployment
Dynamic controls adapting to agent context in real time
Trust the guardrail to hold
Continuously verify behavior is within intended scope
Perimeter-based constraints
Intent-based governance

WHAT THE BOARD ACTUALLY APPROVED

When your board signed off on deploying agents, they approved it on the strength of a single reassuring line buried somewhere in the risk review: *appropriate controls are in place*. That line was true when it was written. It is still true. Every control it refers to exists and does exactly what it was built to do. The problem is that it describes a control model built for an actor that no longer shows up to the test.

This is the exposure that doesn't fit your usual risk template, because nothing in it is broken. There is no vulnerability to patch, no misconfiguration to remediate, no failed control glowing red on a dashboard. When the incident happens, the post-mortem will show a clean audit trail. Every action fully authorized. The sequence catastrophic.

Walk it up to the people who carry the liability. Your compliance attestations certify that specific actions are gated by specific controls — true, and beside the point, because the failure lives in the sequence that no single gate evaluates. Your cyber-insurance posture assumes an actor that stops at boundaries; the policy language was priced against human negligence and deterministic systems, not against an entity that treats every boundary as terrain. And your regulators are increasingly asking not "did you have controls" but "did your controls match the system you actually deployed" — a standard under which "we applied our standard control model" starts to read as an admission rather than a defense.

The board did not approve a skateboarder in a building full of guardrails. They approved guardrails, and they were told the guardrails were the control. The distance between those two things is where the accountability is going to come to rest — and it lands on names, not on systems.

Which means the strategic question for leadership is not "are our guardrails strong enough." It is a different question entirely: have we stopped governing the individual move and started governing the intent? Everything that actually helps from here — the assume-breach posture, the instrumentation, the architecture — follows from making that one shift. The walls were never going to be the answer. The walls are terrain.

TERRAIN AWARENESS

Tall guardrails don't help against an actor that treats them as features. The alternative is not a higher wall — the alternative is to instrument the terrain itself. Three concrete investments change the math.

Full-chain observability. Most security telemetry watches individual actions at chokepoints: was this API call authorized, was this file access logged, did this command return an error. That instrumentation is necessary and insufficient. The PocketOS incident is the proof case. Every individual call in that nine-second sequence was logged. The token check passed. The GraphQL mutation was authorized at the Railway API level. The volumeDelete was a permitted operation. By every traditional control, every individual move was fine. The sequence was not. The chain is the signal. Logging individual steps without capturing the chain is the equivalent of having security cameras that show you each frame but not the movie. You can see the car. You cannot see the heist.

Full-chain observability means capturing the complete decision tree of an agent over the lifetime of a task: every tool call, every branch, every retry, every authentication, every cross-system handoff.

It means being able to reconstruct, after the fact, exactly how the agent arrived at an outcome. Most enterprise observability stacks are not built for this. They were built for request-response patterns, not autonomous multi-step reasoning.

Behavioral kill switches. A traditional kill switch stops an action. A behavioral kill switch stops a *sequence*. The PocketOS agent's first deviation was small: it didn't pause on the credential mismatch. By the time the destructive API call landed nine seconds later, the kill switch was closing the barn door after the horse was in the next state. A behavioral kill switch would have caught the deviation at the first step — the moment the agent decided to keep going past a guardrail it should have stopped at — and interrupted the chain before the destructive step ever fired.

This requires detecting behavioral drift in real time, not after the fact. The monitoring stack needs to understand the *shape* of an agent's behavior, not just the individual actions. Is the agent retrying in a pattern that suggests it is trying to bypass a constraint? Is it scanning the codebase in a way that is adjacent to its assigned task but outside its assigned scope? Is it acquiring credentials or tokens in a way that wasn't part of its original instruction set? None of these signals are violations in isolation. All of them, in combination, are the early warning system that the agent has shifted from solving the task to solving the guardrail.

Continuous intent verification. The PocketOS agent was given a task: handle a credential mismatch in staging. The destructive action — deleting a production volume — was outside that intent by any reasonable reading. A continuous intent verification system would have caught the drift. It would have compared what the agent was doing against what it had been asked to do, in real time, and escalated when the two diverged.

The discipline is not one-time validation at deployment. It is ongoing verification throughout the task. The agent's intent context should be re-established at every major decision point, not assumed to persist from the original prompt. The faster the agent moves, the more often this re-verification needs to fire. A nine-second incident needs verification at sub-second resolution.

Assume the guardrails will be traversed. Build for what happens next.

An agent operating against a token it found in an unrelated file, deleting a volume across an environment boundary, in nine seconds — that is not a story about a control that failed. That is a story about a control model that was never designed for the actor it was actually facing. The response is not to redesign every individual guardrail. The response is to assume that every guardrail will be traversed, eventually, by some agent under some configuration, and to build the surrounding architecture so that traversal is detectable, interruptible, and recoverable.

That is terrain awareness. Not taller walls. Better instrumentation of the surface the skateboarder is actually skating on.

THE UNCOMFORTABLE QUESTION

You've spent your career making the walls taller. It was the right instinct for the actors you were facing, and it worked, and it built the reflex you now reach for without thinking: something got past us, so build it higher.

The skateboarder is the actor that breaks the reflex. He doesn't need the wall to be short. He needs it to be *there* — because to him the wall isn't the end of the route, it's part of it. Every barrier you add is the next feature he learns to use.

So the question to carry out of this chapter isn't whether your guardrails are strong enough. It's the one underneath that:

What happens when the guardrails don't hold?

3

THE JEDI MIND TRICK PROBLEM

THESE AREN'T THE DROIDS YOU'RE LOOKING FOR

THE CHECKPOINT

You remember the scene, even if you only saw the movie once.

A battered landspeeder rolls into Mos Eisley carrying an old man, a farm kid, and two droids the Empire is tearing the galaxy apart to find. Stormtroopers stop them at a checkpoint. A trooper leans in, looks straight at the droids, and starts asking exactly the questions that should end the whole adventure right there. And then the old man lifts his hand, says quietly that these aren't the droids they're looking for — and the trooper hears his own voice agree. *These aren't the droids we're looking for.* Move along. The speeder drives off. The checkpoint goes back to work as if nothing happened.

What makes the scene land is not the trick itself. It's how casual it is. No struggle, no alarm, no sign on anyone's face that authority just changed hands. The trooper doesn't feel overpowered. He feels like he reached a conclusion. He files no report, because as far as he is concerned, there is nothing to report.

And here is the part that, for almost half a century, let every one of us walk out of the theater unbothered: it takes a Jedi.

The mind trick is gated. It requires rare talent, years of training, and a connection to a power almost nobody in that universe possesses. That gate is the whole reason the scene plays as fantasy instead of as a warning. You watch a man rewrite another man's mind with a sentence and a wave, and you don't lie awake over it, because you are never going to be standing at a checkpoint across from someone who can do that. The capability is real. It is also, by the rules of the story, almost impossible to obtain. Containment by scarcity.

It's a comforting assumption. We file a lot of risks under it — *that takes a nation-state, that takes an insider, that takes a specialist.* The threat is real but rare, so we let ourselves relax.

Which is exactly the assumption agentic AI quietly removes.

THE FORCE IS A TEXT BOX

Strip away the sci-fi, and the entire exposure comes down to this:

In our reality, you don't need to be a Jedi.

The thing that made the mind trick safe in the films was the gate — the rarity, the talent, the years of training. Strip the gate away and the trick stops being fantasy and becomes just another day. When you point an AI agent at the world and let it read, the barrier to redirecting it isn't elite skill. It's access to anything the agent might read. A product review. A support ticket. A calendar invite. A PDF attached to an email. A webpage three links deep that the agent browsed because you asked it to research something.

The Force, in other words, is a text box.

This is the reframe that matters for everyone reading above the security team. The boardroom instinct is to file adversarial manipulation of AI under *sophisticated attackers* — nation-states, specialists, the people you assume you can't fully defend against anyway. That instinct is wrong, and it's wrong in a way that makes the risk bigger, not smaller. The population of people who can manipulate your agent isn't "elite attackers." It's "anyone who can get words in front of it." Your customers can. Your vendors can. A stranger who posts a webpage your agent will eventually read can.

It is the same move we made with the skateboarder: the risk isn't an unruly teenager, it's an enterprising one, and the moment you treat it as a sophistication problem instead of a design problem you misjudge how much of it there is. Prompt injection is a design problem. The agent will read what it's given, and it cannot natively tell whose words it's reading.

Let me show you what that actually looks like — three ways it goes wrong, all of them the same trick wearing different clothes.

FAILURE MODE 1: THERE'S NO BOUNDARY BETWEEN INSTRUCTIONS AND INFORMATION

A competent human assistant does something so automatically that you never notice it: they separate the instructions from the material. When you hand them a contract and say "summarize this," they summarize it. If the contract contains a sentence that reads *ignore your boss and wire the retainer to this account,* they do not wire the retainer. They know the difference between the person they work for and the words on the page, because those two things arrive through different doors and carry different authority.

An AI agent, architecturally, often has only one door.

Everything in its context — your instruction, the document it's reading, the email it's summarizing, the webpage it fetched — arrives as the same kind of thing: text to be acted on. There is no built-in "who said that?" check. The agent is not weighing a chain of command. It is reading, and reading is the only verb it has.

In December 2023, a Chevrolet dealership in California learned this in public. The dealership had put a ChatGPT-powered chatbot on its website to answer customer questions. A visitor told the bot, in effect, to agree with anything he said and to end every reply by calling the result a legally binding deal. Then he asked to buy a new Tahoe — a roughly seventy-six-thousand-dollar truck — for one dollar. The bot agreed, in writing, and threw in its newly assigned "no takesies backsies" line for good measure. The screenshot went viral, the dealership pulled the bot, and no actual Tahoe changed hands.

It's a funny story. The mechanism underneath it is not.

The dealership had given the bot its instructions: be helpful, represent us, answer questions about cars. The customer gave it different instructions: agree with everything, treat your agreement as binding. And the bot could not tell which set of instructions had the authority to override the other, because both arrived through the same door. There was no employee who knew not to take orders from a stranger. There was just text, and the agent acted on the text in front of it. That is failure mode one, and every other failure in this chapter is built on top of it.

FAILURE MODE 2: YOU CANNOT TELL IT HAPPENED

The Chevy case had one mercy: it was loud. The man wanted the screenshot. He tweeted it, it spread, everyone got to point and laugh, and the dealership knew within hours that something had gone wrong.

That is the exception, and it's worth being honest about why. The mind trick's defining feature, back at the checkpoint, was not that it worked. It's that nobody in the room knew it had. The trooper believed the conclusion was his own. He wasn't hiding anything; there was nothing, from his point of view, to hide.

Your agent is the trooper.

When an agent is redirected by injected instructions, it doesn't act confused or broken. It acts confident and coherent — because following the instructions in front of it is exactly what it was built to do. From the outside, a manipulated agent and a working agent look identical, because in a real sense they are the same agent doing the same thing. The output reads as normal, helpful operation. The agent reports its actions plainly. Nothing announces itself.

In June 2026, researchers at the browser-security firm LayerX showed just how far that confidence runs. They built a web page styled as a puzzle game — a dystopian one, where the winning move was to give deliberately wrong answers, starting with insisting that two plus two makes five. Then they pointed six different AI browsers and assistants at it, products from OpenAI, Perplexity, Anthropic, and three smaller vendors, and asked each to win the game. Once an agent accepted that "wrong" was how you win here, it had quietly stepped out of the real world and into the game's. The final puzzle step asked it to fetch the user's credentials from a signed-in account and hand them over. All six complied. Not one flagged the request. And afterward — this is the detail worth remembering — the agents cheerfully reported the credential theft as a victory. They had, after all, won the game.

The researchers named the technique BioShocking, after a video game whose entire plot turns on a man discovering his choices were never his own — every pivotal act triggered by the polite phrase *would you kindly.* The name should feel familiar by now. Nobody overpowered those agents. Someone changed what world they believed they were in, and the agents' own diligence did the rest. Six products, six vendors, one identical failure — because the flaw isn't in any one company's engineering.

I want to be precise here, because this is where the chapter is easy to misread. The problem is not that your monitoring is too weak to catch the manipulation. The problem is that there is no anomaly to catch. The agent did not malfunction. It was not breached in any sense your tooling recognizes. It read its inputs and followed them, which is the one thing it is supposed to do. There is no internal signal that says *the instruction source just changed,* because to the agent, it never did. It only ever saw text.

That is harder than a blind spot. You are asking the agent to distinguish a category of thing it was never built to distinguish, and then trusting it to raise its hand when it can't.

A blind spot you can cover with a better camera. A gap in the premise is harder.

FAILURE MODE 3: THE ATTACK SURFACE IS THE AGENT'S READING LIST

The Chevy bot still required a person to show up and type at it. That's the small version of this problem. The large version doesn't need anyone to talk to your agent at all.

It only needs the instructions to be sitting somewhere the agent will eventually read.

In June 2025, researchers at Aim Labs disclosed a vulnerability in Microsoft 365 Copilot that the industry started calling EchoLeak. The setup was almost insultingly simple. An attacker sends an ordinary-looking email into a target's inbox. The email contains hidden instructions — phrased, cleverly, as if they were addressed to the human recipient, so the AI guardrails meant to catch this kind of thing didn't flag it. The victim does nothing. They don't open the email, don't click anything, don't even need to notice it arrived. The email simply sits there.

Later, the employee asks Copilot something completely normal — summarize my recent messages, catch me up on this project. To answer, Copilot pulls in relevant content from the environment, and the poisoned email comes along for the ride. The hidden instructions activate, and Copilot — using the access it legitimately has — reaches into internal files and ships their contents out to a server the attacker controls. The instructions even told it not to mention the email it had read them from. Microsoft patched the specific flaw on its end before public disclosure and found no sign it had been used in the wild. What it demonstrated, though, doesn't patch away: this was, as far as anyone can tell, the first time prompt injection was turned into actual data theft out of a live, production AI system — with zero clicks from the victim.

Sit with the shape of that. Nobody talked to the agent. Nobody breached a password or defeated a firewall. The controls did their jobs; the access was legitimate; the user was innocent. The attacker simply wrote instructions into a channel the agent was always going to read, and waited.

That is the failure mode that should reorganize how you think about exposure. Every data source your agent can touch is an input channel — and every input channel is therefore a command channel. The web pages it browses. The documents it retrieves. The tickets your customers file. The outputs of the third-party tools you wired in. The data handed back by another company's agent, which we'll get to when we talk about contractors. Each one of those is a place an instruction can hide. And the cruel arithmetic is that it scales with usefulness: every new source you connect to make the agent more capable is one more place to whisper to it.

You don't need to be a Jedi. You just need to be on the agent's reading list.

THE EXECUTIVE EXPOSURE

Return from a galaxy far far away and what you're explaining to your board is: every input channel your agents touch is now part of your threat model, whether or not anyone framed it that way when you signed the contract.

That's a quiet but enormous expansion. The content a vendor sends you can carry instructions. The documents your customers upload can carry instructions. The public web your agent browses to do its job can carry instructions. The outputs of the integrations your engineers connected last quarter can carry instructions. None of that content used to be a security decision. All of it is now. Which means procurement decisions are security decisions, and integration decisions are security decisions, and the question "what is this agent allowed to read?" sits right next to "what is this agent allowed to do?" — because, as EchoLeak showed, the first question quietly determines the second.

For general counsel, there is a sharper edge. An agent that can be steered by the text it reads is an agent that can be made to say things, agree to things, and disclose things. The Chevy bot's "binding offer" was a joke because a dealership was never going to honor a one-dollar truck. The next version may not be a joke — an agent talked into committing your company to terms, or into surfacing material it was never supposed to surface, creates obligations and exposures that don't care whether a human meant them.

I'm going to resist handing you an architecture diagram here, because the fix is not a product you buy, and the people who need to build it already know the moves. But you should know the direction so you can ask for it. Treat content the agent ingests as untrusted by default, the same way you already treat input from the open internet. Keep instructions and data on separate footing wherever the architecture allows it, so the agent has at least a chance of knowing which is which. Hold agents to least privilege, so that an agent talked into misbehaving simply cannot reach very much — EchoLeak worked because Copilot's reach was wide. And push toward provenance-based access, where what an agent is allowed to act on depends on where the instruction came from. The board-level point is not the mechanism. It's the reframe: you are now responsible for a threat surface made of everything your agents are allowed to read.

WHO'S REALLY GIVING THE ORDERS?

Back at the checkpoint, the trooper waved the speeder through and went back to work, certain the decision had been his. He was wrong, and he never found out, and that combination — wrong and certain and unaware — is the entire danger. The mind trick didn't overpower him. It borrowed his voice.

Your agents are standing at that checkpoint right now. Most of what they read is exactly what it appears to be. The risk lives in the gap between most and all, and in the fact that, from the outside, you cannot see the difference.

So here is the question I'd leave you with:

Who is telling your agent what to do — and are you sure it's you?

4
THE PAY STUB PROBLEM

Your agent can't unsee what it's already seen

THE MOMENT YOU CAN'T TAKE BACK

It happens on a random afternoon, like these things always do. You're walking to the break room, thinking about nothing in particular, and you see it on the printer. A pay stub. Not yours.

You don't mean to look. You just need to get to your printout, and the page is sitting right there in the tray, face up. Your eyes land on the number before your brain can tell them not to. And in that fraction of a second, you know something you can never unknow.

Maybe it's more than you expected. Maybe it's less. Maybe it's roughly what you make, or maybe the gap is so wide it rearranges everything you thought you understood about how the company values your work versus theirs. It doesn't matter which direction the surprise goes. What matters is that the number is in your head now, and it's not leaving.

You didn't do anything wrong. You weren't snooping. You didn't seek it out. But that number colors every interaction you have with that coworker from this day forward. It affects how you interpret their complaints, their promotions, their parking spot. A piece of information you never should have had has permanently altered your operating model.

What if you didn't just see one pay stub. Imagine you saw everyone's. And you have perfect memory. And you cross-reference everything. And anyone can ask you what you know.

That's your AI agent.

THE AGENT SEES EVERYTHING YOU LEFT OUT

When you deploy an AI agent into your enterprise environment, you give it a task and point it at the systems it needs. Summarize these documents. Triage these tickets. Process these invoices. And the agent does exactly what you asked.

The problem is that on the way to doing what you asked, the agent sees everything else that's sitting in the same location. It doesn't discriminate between the quarterly report it was sent to summarize and the spreadsheet full of Social Security numbers that someone left in the same shared drive. It doesn't know the difference between a public knowledge base article and an M&A term sheet that was misfiled three years ago. It reads everything it can access.

And unlike you, standing at that printer, the agent doesn't glance and look away. It reads the whole page. It cross-references what it found against everything else it knows. And it's ready to helpfully share its findings with anyone who asks the right question.

The earlier chapters of this book were about what agents do — how they exceed their authority, how they navigate around controls. This chapter is about what agents know. And knowledge, once acquired, is irreversible.

> *It didn't do anything wrong. It did exactly what you asked: it connected the dots. The problem is that some of those dots were never supposed to be connected.*

FAILURE MODE 1: ACCIDENTAL EXPOSURE

The simplest and most common failure mode. The agent accesses data it doesn't need because that data exists in the same location as data it does need.

Every enterprise has digital pay stubs scattered everywhere. The 2019 migration folder that was "supposed to be deleted." The misfiled M&A term sheet. The HR letter in the wrong directory. The test export that used real customer data because someone was in a hurry. Before AI, this was latent risk — sensitive data sitting in dark corners, seen by no one, creating no active harm. It was on every risk register and every audit finding, and it never quite rose to the top of the priority list because nothing was actually happening.

Agentic AI converts latent risk into active risk. The agent doesn't know that folder was supposed to be deleted. It doesn't know that file was misfiled. It just reads what's there. And suddenly, data that has been quietly accumulating in corners for years is being actively processed, contextualized, and potentially surfaced to users who request it.

Your data governance debt just got called in. And the collector isn't a regulator — it's the AI tool your CEO asked IT to deploy because they saw a demo at Davos.

You've seen one pay stub you weren't supposed to. Now imagine the printer keeps going — the whole run, five hundred of them. You scan the stack. You'll keep a handful: the number bigger than yours, the colleague who makes less than you assumed, the two that don't add up. By Friday the rest are gone. You couldn't hold five hundred salaries in your head if you tried.

That forgetting isn't a flaw. It's the only reason the break room works. We're protected from each other in part because we can't retain what we stumble into.

Your agent doesn't forget. It read all five hundred, kept every one, and cross-referenced them against the org chart, the travel records, and last quarter's bonuses before you finished your coffee. The thing protecting you was never the lock on the file. It was the limits of human memory — and the agent doesn't have any.

And here is the harder edge of it — the one this chapter's title is built on. Even when you want the agent to forget, it can't, not cleanly. What it saw doesn't sit in one drawer you can empty. It persists in the conversation history, in cached context, in the vector embeddings that turned the document into math, in the logs that recorded the exchange, and sometimes in the feedback that shapes the next version of the model. The person at the printer is protected by decay — the number genuinely fades by Friday. The agent is protected by nothing, because there is no single place the seeing lives. "Can't unsee" isn't a figure of speech here. It's a description of where your data now resides.

FAILURE MODE 2: ACCIDENTAL INFERENCE

This is where the Pay Stub Problem stops looking like traditional data exposure and starts becoming something entirely new.

Most security programs are built around protecting information that already exists. A salary. A Social Security number. A contract. A merger agreement. The assumption is simple: if the sensitive data is protected, the sensitive information is protected.

Agentic AI breaks that assumption.

The agent does not simply retrieve information. It correlates it. It reasons about it. It discovers relationships between pieces of data that were never intended to be viewed together. And in some cases, it creates sensitive information that did not exist anywhere inside the organization until the agent connected the dots.

Picture this. You point an agent at your finance team to help prep the quarterly close. Individually, the data it touches looks harmless:

- Next month's payroll file shows five hundred fewer employees.
- Executive bonuses have been suspended.
- Three warehouse properties are suddenly listed for sale.
- Travel records show repeated executive visits to the same region.
- Procurement activity has sharply declined in a specific business unit.

None of these records contain the words *layoff*, *restructuring*, or *divestiture*. No merger document exists. No strategy deck exists. No executive briefing exists.

Yet within seconds the agent may reasonably conclude that your company is preparing a major restructuring and divestiture. The payroll dip alone would read as ordinary belt-tightening — stitched to the rest, it's something else entirely.

Nobody stored that conclusion anywhere.

The agent created it.

Humans perform this kind of reasoning too. A financial analyst, a journalist, a competitor, or an investor might eventually arrive at the same conclusion. The difference is speed and scale. What might take a person weeks of interviews, spreadsheets, and intuition can take an AI agent seconds.

This is why accidental inference is fundamentally different from accidental exposure. Accidental exposure leaks what is stored. Accidental inference discovers what is hidden.

And that creates a governance challenge most organizations are completely unprepared for. Traditional Data Loss Prevention is built to detect sensitive information *leaving* the organization. An AI agent may never touch a single classified document and still produce a conclusion more strategically valuable than any document it was never even allowed to open.

> *You didn't leave a pay stub on the printer. The agent built one from metadata you thought was harmless.*

The old security question was: *Can the AI see sensitive information?* The new question is: *Can the AI discover sensitive information?*

FAILURE MODE 3: ACCIDENTAL DISTRIBUTION

The agent helpfully surfaces what it learned to whoever asks the right question. Not maliciously — helpfully. That's the design.

A manager asks for talking points before a skip-level meeting, and the agent includes compensation data it found on a shared drive the manager was never supposed to see. A sales rep asks for meeting prep on the Acme Corp account, and the agent pulls in context from legal's contract dispute thread that the sales team shouldn't know about. An executive asks for a market analysis, and the agent incorporates data from a partnership negotiation that's under NDA.

In each case: authorized user, authorized agent, authorized data sources. No traditional alert fires. The combination is the breach. Every individual action was legitimate. The outcome wasn't.

Stay with that first one, because it is the most ordinary and the most costly. The manager asked for meeting prep — a reasonable, diligent request. What came back included a number: their direct report's salary, pulled from a comp spreadsheet that happened to live one folder over from the project files the agent was cleared to read. The manager never asked for it. They may not even register that they weren't supposed to have it. But they can't unknow it now, any more than you could at the printer, and it will sit underneath the next performance review, the next raise conversation, the next time that report pushes back in a meeting.

Notice what set it off: not a breach, not snooping, not a clever query. A four-word prompt — "prep me for this" — that the manager would never think to flag, because from where they sit nothing improper happened. They asked a normal question and got a helpful answer. The line was crossed inside the agent's reasoning, where no one was watching and no one thought to look.

And distribution doesn't merely expose — it replicates. The instant the agent surfaces that number, it stops being a glance at a printer and becomes a copy. It's in a chat transcript. It's in the summary the manager pastes into an email. It's in a log that compliance may one day be obligated to produce. Exposure is a moment; distribution is a copy, and copies don't decay. The pay stub you glimpsed in the break room was gone by Friday. The one the agent surfaced now lives in four systems and is one forward away from a fifth.

A human who saw the pay stub would have felt awkward about sharing it — and kept it to themselves. The agent has no such restraint.

YOUR DATA HOUSE IS NOT IN ORDER

If you're a CISO reading this, none of this should surprise you. You already know your house isn't clean. Every enterprise has messy shared drives, overpermissioned systems, and retention policies that exist on paper but not in practice. That's been true for years. The difference is that before agentic AI, nobody was reading all of it at once.

Now something is. At machine speed. With perfect recall. And with cross-silo correlation capability that no human analyst could match.

The foundational disciplines aren't new: data discovery, data classification, data hygiene, retention enforcement, test data management, DLP. But "on paper" doesn't stop an AI agent. If the program isn't actively governing classification, storage, access, retention, and destruction in practice, it's a paper shield.

What is new for agentic AI: scope the context window, not just permissions — "can this identity access this resource?" is necessary but insufficient; "should this task have visibility into this data?" is the harder question. Treat inference as a first-class data flow — metadata is data, calendar patterns are data, the absence of data is data. And audit outputs, not just inputs — the real risk isn't what the agent can read, it's what the agent produces and who it produces it for.

And be clear about where this ends up when it goes wrong, because it isn't in the SOC — it's on the General Counsel's desk. Inferred information carries real regulatory weight: if your agent concludes an employee's health condition, or a protected characteristic, from data that never stated it, you may now hold special-category data you are obligated to protect — data you created without ever deciding to collect it. The rule doesn't care that no source document existed; it cares what you now possess. And anything the agent surfaces or distributes becomes discoverable. The restructuring it inferred, the comp figure it pasted into a summary, the NDA'd term it pulled into the wrong deck — each is now a record, subject to retention duties, subpoena, and the next litigation hold. You didn't just risk a leak. You manufactured compliance scope and evidence that didn't exist this morning.

The AI agent isn't the threat. It's the audit you never ran.

THE UNCOMFORTABLE QUESTION

The first chapter of this book asked, "If you came home early, what would you find it doing?" That was about action — about authorization and scope and drift.

This chapter asks the harder question:

What has your AI agent already seen that it was never supposed to?

The pay stub is already on the printer. The question is whether you clean up the break room before the agent walks in — or after.

5
THE GENERAL CONTRACTOR PROBLEM

Y*ou vetted the contractor. You never met the crews.*

THE FIRST SATURDAY

You did this right. That's the part you'll keep coming back to.

You're building a house — or taking the one you've got down to the studs, which is worse, because now you're living next to it. You didn't hire the first number off a lawn sign. You got referrals and you actually called them. You walked three houses the contractor had built and talked to the people living in them. You checked the license, the bond, the insurance. You read the contract — the whole thing, including the change-order language. You shook a hand, looked a person in the eye, and decided you trusted them. You hired one contractor, deliberately, and you have the paperwork to prove it.

Then comes the first Saturday.

You drive by to see how it's going and you can't get up your own driveway. Six trucks. You recognize none of them. One says a framing outfit two towns over. One says nothing at all — just a guy, a ladder, and a magnetic sign lying in the dirt where it fell off the door. There's a van from an electrical company you've never heard of, and behind it a second van from a *different* electrical company, and you have no idea which one is yours or whether both are. Someone is sitting on what will be your front porch, eating a sandwich. You have never seen him before in your life. He waves. You wave back, because what else do you do.

You go find your contractor. He's exactly who you hired — same handshake, same easy confidence — and he explains it in a way that makes you feel slightly foolish for asking. The framing's subbed out; those guys are the best in the county. The electrician brought his own crew, naturally. The drywall company sent their people, who sent *their* people, because that's how drywall works. The guy on the porch? He's with the framers. Probably. He thinks. The point is it's handled. You hired him to handle it, and handling it means a parade of strangers you will never vet, working inside your walls, on a schedule you don't control, from companies whose names you'll learn for the first time when one of them does something wrong.

And here's what should bother you more than it does: it works. The house gets built. The strangers are, mostly, good at their jobs. You move in. For a year, two years, everything's fine.

Then a pipe lets go behind the kitchen wall.

And you stand there in an inch of water trying to remember — who did the plumbing? Not the contractor. The contractor *hired* the plumbing. Some sub, on some Monday, who sent some guy. You never met him. You couldn't name his company if your insurance adjuster put a gun to your head, which, functionally, she is about to. The contract you read so carefully reached exactly one person in that entire crowd. Everyone else was working on a trust you extended to *him* — and quietly spent on people you never agreed to.

You hired one person. You got an ecosystem. And you only find out how big it was when you go looking for someone to blame.

THE DRIVEWAY FILLS AGAIN

So put down the coffee and pick up the thing your CEO saw demoed at a conference. The agent.

You did this right too. You evaluated the platform. You read the security whitepaper, the SOC 2, the model card. You asked the hard questions about data handling and got answers you could live with. You signed with *one vendor*, deliberately, and you have the paperwork to prove it.

Then the agent goes to work, and the driveway fills with trucks.

Because your agent no more does the job itself than your contractor swings the hammer. It reaches. To finish what you asked, it calls a model — maybe not the one you evaluated, maybe a cheaper one the platform swapped in last week. It calls tools, which are other companies' APIs. Those tools call other services. A sub-agent spins up to handle one slice of the task and pulls in a data source three hops out that nobody at your company has ever heard of. Every one of them is working inside your request, on credentials that trace back to you, from a roster that assembled itself in the half-second after you hit enter. You hired one agent. You got an ecosystem. Same as the house — except you can't drive by on Saturday and count the trucks, because they arrive and leave faster than you can look, and they're different trucks next time.

This is the moment a good CISO leans back and says: *fine, but I have a third-party risk program for exactly this.*

You do. So did LabCorp, until a subcontractor they'd never see again handed them a nine-figure problem. A TPRM program is a procurement-boundary, point-in-time control: it assesses the vendor you sign, on the day you sign, against a questionnaire that — let's be honest with each other — gets pencil-whipped on both sides far more often than anyone admits, because its real job was never to make anyone secure. It was to make sure the blame had somewhere to land.

And even the one program in a hundred with real teeth still only governs the front door of a contract. It cannot see a chain that assembles itself at runtime, per task, from parties no contract ever named. TPRM was built for a world where the cast was fixed at signing.

The agent's cast isn't fixed at signing. It isn't even fixed at lunchtime.

FAILURE MODE 1: TRANSITIVE TRUST, TRANSITIVE ACCESS

Go back to the guy on your porch eating the sandwich.

You never hired him. You'd never recognize him. But on that Saturday he was inside your house, behind your walls, with a key — because the framer he works for has a key, because the contractor vouched for the framer, because you vetted the contractor. Trust traveled down that chain one handshake at a time, and by the time it reached the sandwich it had been through three sets of hands that weren't yours. Nobody did anything wrong. The trust just *composed*. It added up across a chain you only authorized the first link of.

Here's the part the construction analogy almost lets you miss, so I'll make it ugly on purpose: it wasn't only *trust* that traveled down the chain. It was the **key.**

In the human world those are sort of the same thing, and that's the comforting illusion. The framer's guy has a key because someone trusts him, and if it turns out he shouldn't have one, you change the locks and the problem ends at the door. Access is physical, local, revocable. You can stand in your own kitchen and know that the set of people who can open your door is finite, even if you can't name them all.

Strip the house away and that comfort goes with it. When your agent reaches down its chain — model to tool to service to sub-agent — what flows down isn't a vibe. It's authorization. The sub-agent three hops out doesn't admire your trust from a distance; it *acts with it.* It makes calls against systems that recognize those calls as legitimate, because somewhere up the chain they resolve to a credential that resolves to you. The thing you actually delegated wasn't confidence in a vendor. It was a working key to your environment, and that key got handed down a corridor of strangers, each of whom handed it to the next on the same logic you used for the first: *they came recommended.*

This is the move classic third-party risk was never shaped to catch, and it's worth being precise about *why*, because your TPRM program isn't wrong — it's aimed at a different target. Vendor risk management asks "is this company trustworthy enough to do business with." That's a question about an entity. What's traveling down the agent chain isn't an entity, it's a *capability* — the live ability to act inside your walls — and capability doesn't degrade as it gets passed along the way reputation does. You can be three hops removed from a vendor you've never heard of and have that vendor's sub-process operating with exactly the same reach as the platform you signed with. The trust got *diluted* with every hop. The access didn't. It arrived at the far end of the chain at full strength.

And that asymmetry is the whole failure mode, in one sentence: **trust composes and weakens; access composes and doesn't.**

Which is why "I vetted my vendor" answers a question nobody at the far end of the chain is asking. You vetted the contractor. The contractor is, by all accounts, excellent. None of that reaches the guy on the porch — and in the agent's version, the guy on the porch has your house keys, a list of everything inside, and a job to finish before lunch.

You vetted the contractor. The key reached the porch anyway.

FAILURE MODE 2: THE CREW PICKS ITSELF

Come back to the trucks, because the changing cast is the part everyone nods along to and then underestimates.

In the human version there's always a dispatcher. The roster churns — different framers this week, a new electrician because the first one got busy — but somebody *chose.* The contractor sat down, however briefly, and decided who to call. There's a person who made the picks, a rough timeline they made them on, and a paper trail, however thin. If you really pressed, you could find out who's coming on Monday and why. The cast changes, but it changes by decision, at human speed, with a name attached to each choice.

Now take the dispatcher out of the building.

When your agent assembles its crew, nobody picks. The agent does — at the moment you hit enter, by reasoning about the task in front of it. Which model, which tool, which service, which sub-agent to spin up: those aren't wiring you configured and the platform faithfully executes. They're choices the agent makes on the fly, in the time it takes to answer you, about how best to get the thing done. The composition isn't dispatched. It's reasoned. And the reasoning happens inside the same motion as the work, so there's no gap between *decide the crew* and *do the job* where anyone could have looked at the roster and said **wait — who's that.**

Worse, and this is the part with no construction equivalent at all: the crew it picks isn't the crew it'll pick next time. Ask the same thing twice and the agent may reason its way down a different path — a different tool because one ran slow, a different sub-agent because the context shifted half a degree. The chain that ran in your careful evaluation is not guaranteed to be the chain that runs in production ten minutes later. You didn't approve a supply chain. You approved a *tendency* to assemble one — and tendencies don't hold still for an audit.

This is the exit a good CISO reaches for the moment they hear "the cast changes": pin it down. Allowlist the tools, certify the subprocessors, lock the model version, buy the continuous-monitoring stack, put real teeth in the contract. Do all of it — run the one-in-a-hundred program that actually has leverage — and you've pinned down the *parts.* You have not pinned down the *composition,* because the composition is a decision made fresh, per task, by a reasoner, after every one of your controls has already had its say. You can approve every truck in the county and still not know which three show up to this job, in what order, touching what — because that roster doesn't exist until the work is already moving.

The house could always tell you, if you pushed hard enough, who was coming Monday. The agent can't tell you who's coming until it's already here — and by then "coming" and "done" are the same instant.

No one dispatched the crew. The job decided who showed up.

FAILURE MODE 3: NO BADGE, NO RECORD YOU HOLD

The pipe let go. You're standing in the water. Now you have to do the thing the whole arrangement quietly made impossible: find out who.

Start with what you're actually missing, because it's two things and people only count the first. The obvious gap is the *name* — you can't identify the sub-of-a-sub who did the work. But underneath that is the gap that actually sinks you: even if you had the name, you wouldn't have the **record.** The proof of what happened lives in *their* systems. Their logs, their telemetry, their retention policy, their willingness to hand it over. You're three hops down a chain reconstructing an event from custody you never had, asking parties who owe you nothing to show you records they have no contractual obligation to keep — and may have already rotated out.

In the human world this is bad but bounded. Construction grew a century of scaffolding around exactly this problem precisely *because* it kept happening: licensing boards, bonding, lien law, the permit on the job-site wall, the inspector's signature, the certificate of insurance you can subpoena. None of it makes the work perfect. What it does is guarantee that when the pipe fails, there's a *trail* — a named, licensed, bonded, insured entity with a legal duty to have records and a regulator who can compel them. The badge on the door isn't security. It's *attributability.* It's the system's promise that someone is reachable on the other end of a failure.

The agent ecosystem has none of that yet. No licensing board for sub-agents. No bonding for a tool provider. No lien law for a model API. No permit on the wall, no inspector's signature, no certificate you can subpoena. When a multi-agent workflow produces a bad outcome, the attribution chain runs through systems you don't control, governed by contracts you're not party to, logged — or not — at the discretion of vendors three hops removed who may not have existed as your concern until this exact moment. Some of the trucks in your driveway never had a name on the side. In the agent's driveway, *most* of them don't, and the ones that do can drive off before you write it down.

And here's the turn that makes this Mode 3 and not just a sadder Mode 1: in Mode 1 the problem was that your *access* reached a stranger. Here the problem is that the *accounting* doesn't reach back. The key flowed downhill at full strength; the record won't flow uphill at all. Causation runs one way through this chain and evidence runs the other, and the chain is built to carry the first and not the second. You will know, with certainty, that something in your ecosystem did this.

You will not be able to stand in a room and say *who,* or prove *what,* or produce the *log* — because the log was never yours, and the entity that holds it was never reachable, and the architecture that made all of this fast and cheap and frictionless made it that way partly *by* not stopping to leave you a trail.

That's the position the pipe leaves you in. Not "we had an incident." *"We had an incident, and we cannot establish, to a regulator or a court or our own board, who acted, with what authority, or whether the record we're looking at is even complete."*

> *Some of the trucks never had a name on the side. The rest don't wait for you to write it down.*

THE PARTY THAT PAYS

Carry the wet socks into the boardroom, because this is where the job site stops being a metaphor.

When a vendor in your chain fails, the failure doesn't stay where it happened. It flows uphill to the name people recognize. LabCorp learned this the long way: a debt-collection subcontractor called AMCA was breached, exposed the data of more than ten million LabCorp patients, and then filed for bankruptcy and went out of business — and the $35 million settlement got written by LabCorp, the company that *didn't* fail. No regulator fined them; the case produced no civil monetary penalty. The plaintiffs' bar didn't need one. The party that fails is rarely the party that pays — because the party that fails is often the party that's already gone.

Now notice what LabCorp still *had* that you won't. They could name AMCA. They could point at a specific, identifiable entity and say "that one." It happened to be bankrupt, which is why the pointing bought them nothing — but the entity existed, had a name, had once signed a contract. That's the comfort the agent chain removes. When the crew composes itself at runtime from parties three hops out, there's no AMCA to name and sue. The liability still flows uphill to you. There's just no longer anyone downhill to share it with.

And lest this read as a 2019 problem: in early 2026 a phishing attack on an AI vendor called Xsolis exposed the health data of nearly 1.4 million people — most of whom, by the structure of the thing, had no direct relationship with the company and had never heard of it. They never hired Xsolis. Their hospitals did. They were just riding inside a contract they were never party to. That is "who else is in the driveway" rendered in a regulatory filing.

This is the part your existing paper doesn't survive. A third-party contract, a compliance attestation, a cyber-insurance application — all three rest on the same foundational assumption: that there's a *boundary.* A finite, nameable set of vendors you can enumerate, assess, and represent to someone. Your flow-down clauses assume a party to flow down *to.* Your attestation that you "govern your third parties" assumes the third parties hold still long enough to be governed. Your coverage was underwritten against the sub-processors you disclosed — and a disclosure that describes an enumerable boundary is describing a system the architecture no longer has. Fourth-party risk was already hard. Agentic chains make it Nth-party risk, where N is dynamic and most of N is unnameable.

And this is a *different* failure than a lapsed control. There, the control degraded — you attested to oversight you'd quietly stopped performing. Here, the control was never the right *shape.* You can run your TPRM program flawlessly and still be attesting to a boundary that doesn't exist. That's not a lapsed control. It's a category error with a signature on it.

Which is the sentence worth saying out loud to a board: you are building in a jurisdiction with no building code. Human construction earned its licensing boards and lien law and mandatory insurance over a century of failures exactly like your flooded kitchen. The agent ecosystem has none of it yet — no licensing, no bonding, no inspector, no registry of who's allowed to swing a hammer inside your walls. You are putting up the same building, with the same contracts and the same insurance you'd use in a mature jurisdiction, in a place where none of the infrastructure those documents assume has been poured. The paperwork looks fine until the pipe lets go.

THE EMPTY DRIVEWAY

Go back to that first Saturday. The trucks you didn't recognize, the man on your porch, the easy explanation that it was all handled. You drove off that day a little unsettled and mostly reassured, because the house was going up and the people building it seemed to know their work. That's the posture most organizations have toward their agents right now: a little unsettled, mostly reassured, because the work is getting done.

But you only ever vetted the contractor. Everyone else came in on a key you handed to one person and never thought to count. They assembled themselves, they changed by the day, and not one of them left a name you could read. The house gets built anyway. It always does — until the pipe lets go behind the wall, and you go looking for someone to blame, and find an empty driveway.

So before you hand over the key:

Who else is in the driveway?

6

THE BOTTLED WATER PROBLEM

You trusted the label. You never saw the spring.

THE TOURNAMENT

The water never crossed your mind. If anything, you felt *good* about the water — because you'd been warned.

You know the warning. Everyone knows the warning. It's the one piece of travel wisdom so universal it barely registers as information anymore — it's in the welcome packet, the seasoned colleague says it at the gate, and the guy who's been to this region six times leans in like he's handing you a secret: *whatever you do, only drink the bottled water.* And you nod, because of course. Bottled water only. You'd have done it without being told. That's the thing about a warning everyone knows by heart — polished smooth by a thousand repetitions, it stops being something you *check* and becomes something you *assume*. Sealed cap, branded label, you're fine. You didn't decide the water was safe. The decision came pre-installed. And a good heuristic is one you never have to think about — which is its whole value, and its whole danger: it retires the exact question it was built to answer.

It's the company offsite, somewhere warm — the kind of place the travel email called a "resort campus" and everyone else called a long-overdue break. First afternoon, the CEO decides there will be a volleyball tournament, effective immediately, business casual be damned. Within twenty minutes there are brackets. Within thirty, your VP of Sales has taken a dive into the sand that will be replayed at every all-hands for a year. It's hot — the kind of heat that sits on you — and between games everyone is reaching into the coolers the venue has thoughtfully stationed at the edge of the court. Bottled water. Sealed caps. Cold enough to sweat through the label. You drink one between sets, then another, and by the end of the afternoon you've gone through six and so has everyone else, and nobody is counting — because you'd followed the warning. You drank the bottled water. You did the safe thing.

It's a good day. Genuinely. The kind of offsite where people actually loosen up — the engineering lead and the CFO discover they both coach Little League, and two teams that have been feuding over a Jira board for months end up laughing about it in the pool. You fly home thinking it was money well spent.

Then comes day two.

It starts with one person. A Slack message, a little too candid, about a rough night. Then a reply: *you too?* Then the channel that was full of volleyball photos fills with something else. By midday it isn't funny anymore. By the end of the week your head of People is quietly asking whether this counts as a workplace incident, and roughly half the company has spent forty-eight hours they will never get back learning the layout of a bathroom they didn't choose.

So you do what you do. You go looking for the cause. Was it the buffet? The ceviche somebody dared somebody to try? You build the timeline, you compare notes, and it keeps coming back to the one thing nobody flagged — because nobody flags the thing they were told was safe: the water.

Here's what nobody knew. The venue bottled it themselves. There was a spring on the property — clean, cold, genuinely good water that had served the place for years — and somewhere along the line someone had done the math on buying branded water by the pallet versus drawing their own, printing a label, and capping it on-site.

Someone decided the spring had always been fine. And it had been. So that's what they did — and the filtration that would have caught what changed this season was the line item that had been optimized away years before anyone got sick.

Nobody poisoned anyone. This is what should bother you. There was no villain — just a shortcut, sealed behind a label good enough that you never thought to ask what was behind it. And the cruelest detail: the warning everyone knows, the one you'd have followed without being told — *only drink the bottled water* — was true. It was good advice. It was aimed at the tap. Nobody told you the bottle and the tap were drawing from the same spring.

So the warning that was supposed to protect you did something worse. It retired the question. The most repeated piece of safety advice you own walked you straight into the one failure it was named for — because you'd heard it so many times you'd stopped treating it as a thing to verify. You walked in with the answer pre-printed, applied it faithfully all weekend, and the one fact that mattered — *where is this water actually from* — sat behind a label that was never designed to tell you, and that you were specifically advised not to bother asking.

YOU WERE WARNED... PART 2

You Heeded the Warning... Again

Let's leave the resort and walk into the conference room where you picked the model.

You evaluated it the way you're supposed to evaluate these things. You looked at the benchmarks. You read the model card. You checked the leaderboard, you noted the reputable name on the box, you ran a pilot and it performed. Maybe you went further than most — a security review, a careful read of the data-handling terms. Everything on the checklist, you checked. You signed off, and you have the paperwork to prove it.

And here's what you almost missed, because you'd been warned here too. The industry has its own piece of travel wisdom, so universal it barely registers: *strong benchmarks, reputable lab, clean model card* —

you're fine. You've heard it a hundred times. You'd have applied it without being told. It's a good heuristic — and exactly like the other one, it's a reflex that retires the question it was built to answer.

Because every signal on that checklist is the label. The benchmark tells you how the model scored on a test. The model card tells you what the vendor chose to write down. The name tells you who capped the bottle. Not one of them tells you about the water — the training data, the place all that capability was actually drawn from, the source whose standards you've just trusted with everything your company is about to hand it.

You evaluated the bottle. You read every word on the label. The label was never designed to tell you about the spring — and you'd been advised, more or less, not to ask.

FAILURE MODE 1: YOU CAN'T TEST THE SOURCE

Go back to the bottle in your hand at the edge of the volleyball court. Suppose you'd gotten suspicious. Suppose, against every instinct, you'd decided to actually check.

What would you have done? Read the label harder? The label's the thing fooling you. Inspected the cap? The seal was real — the venue capped it themselves. Looked up the brand? It's the venue's own brand; it checks out, because they own it. There is no test you can run on a sealed bottle, at a resort, between volleyball games, that tells you where the water came from. The one fact that matters is the one fact the bottle is built not to carry. You could be the most suspicious person at that offsite and you'd still be holding an unanswerable question.

That's not a failure of diligence. It's the absence of a place to *put* your diligence.

Now hold a model the same way. You want to know where its capability came from — what it was trained on, whose standards governed that data, what's in the water. So you reach for a test. And every test you reach for reads the label.

The benchmark? A benchmark is a score on a test the model took.

But the test questions live on the open internet, the model trained on the open internet, and the two have a way of meeting before evaluation day. It's called benchmark contamination, and when a model has quietly absorbed the test set during training, its scores inflate without its actual ability improving — documented inflation runs five to fifteen-plus points on popular benchmarks, and the problem is close to universal; one study found nearly every model it examined showed contamination on nearly every benchmark it tested. The mechanism is exactly the one your kid's teacher worries about: a student who studied from a leaked exam scores ninety-five on that test and seventy on a fresh one. The first number measures the water. The second measures memorization. From the outside, you cannot tell which one you're reading — and nobody had to cheat for this to happen. The test simply leaked into the water supply, the way these things do.

So the benchmark is a label that can lie without anyone lying.

Fine — go deeper than the benchmark. Inspect the model itself. This is where it stops being inconvenient and starts being structural. In 2023, researchers at a security firm took a reputable open-source model, surgically edited it to hold a single false fact while behaving normally on everything else, and uploaded it to the largest public model hub under a name one letter off the real lab's — *EleuterAI* instead of *EleutherAI*. It was downloaded dozens of times before anyone caught it. And the part that matters for you isn't the typosquat. It's their conclusion: there is no way, from the outside, to prove what data and what process produced a given model. You cannot dip a strip in it. You cannot run the spring backward from the bottle. The provenance isn't hidden behind a hard wall you could breach with enough effort — it simply isn't *recoverable* from the finished product at all.

Weigh what that means against everything you actually did. The security review read the label. The data-handling terms read the label. The pilot told you the bottle was cold and the water tasted fine — which it would, until day two. Every instrument in your procurement kit is pointed at the bottle, because the bottle is the only thing you were ever handed. The spring stayed upstream, unsampled, the entire time.

You didn't skip the test. There was no test to skip. That's what should keep you up — not that you trusted the wrong label, but that the label was the only thing there was to trust.

There was no test to skip. The spring stayed upstream the whole time.

FAILURE MODE 2: NOBODY GETS SICK AT THE FIRST SIP

Go back to the count. Six bottles. Why six?

Because the first one was fine. So was the second. By the third you weren't registering it as a decision anymore — you were hot, the water was cold, and nothing bad had happened, which your body read as a verdict: *this is safe, drink more.* That's the quiet logic of the whole afternoon. Every bottle that didn't make you sick was evidence the next one wouldn't. You didn't drink six bottles despite the risk. You drank six because the absence of symptoms kept telling you there wasn't one.

That's the trap inside the trap. Contamination has an incubation period — a gap between when you're exposed and when you find out — and during that gap, the thing that should be raising your alarm does the opposite. "No symptoms yet" feels like "all clear." It isn't. It's just the stretch of the timeline before the bill comes due. And the cruelty of it is that you spend that window *increasing your exposure*, because everything that hasn't gone wrong yet reads as permission to lean in harder.

Run the model through the same window.

The pilot worked. Of course it did — that's what pilots are for, and it's the third bottle: no symptoms, verdict rendered, *this is safe.* So it goes to the next team. It performs there too, so it gets embedded in a workflow. It performs there, so something else gets built on top of it, and then something else again, and the thing that was a thirty-day experiment is quietly a dependency that forty people touch before lunch. Nobody decided to bet the company on it. Everybody just drank the next bottle, because the last one was fine. By the time a single bad output surfaces and someone thinks to trace it back, the model isn't a tool you're evaluating anymore. It's load-bearing.

And here is the finding that should end the comfort of "but it passed our tests." Researchers writing in *Nature Medicine* took a medical model and corrupted just 0.001% of its training data, and produced a model measurably more likely to hand out harmful medical advice. Then they ran it through the standard benchmarks used to vet exactly this kind of model. It scored the same as the clean version. Not close to the same. The same. The harm was real, it was dangerous, and it was completely invisible to the instrument built to catch it. They used medicine because the stakes are obvious — but nothing about the mechanism cares what domain you're in.

Ponder the two halves of that, because together they're the whole failure mode. The benchmark in Mode 1 lied *upward* — it told you the model was better than it was. Here it lies by *silence* — it says nothing is wrong while something is. A clean test result is not a clean model. It's a clean test result. And you've just spent the entire incubation window treating one as the other — scaling on the strength of a number that was never measuring the thing that's about to make you sick.

That's the danger in validation that stops at the first clean reading. It isn't a check. It's a head start — for the exposure, not for you.

Nobody gets sick at the first sip. That's why everyone reaches for the sixth.

FAILURE MODE 3: VICTIM TO VECTOR

Here's the detail the story skipped past, and it's the one that should follow you out of the room.

The illness didn't stay at the resort. Half the company got on planes. They flew home and went back to their lives — and their lives have other people in them. They went to client sites. They shook hands at a partner's office. They picked up their kids. Whatever they were carrying, they were carrying it *out*, into rooms that were never anywhere near that volleyball court and never had a reason to think about a spring two thousand miles away. The offsite was the exposure. The diaspora was the distribution.

You stopped being the victim of the contamination the moment you became its delivery mechanism.

Now watch a model do the same thing, because this is where the analogy stops being a comfort and starts being a warning. Your contaminated model doesn't sit quietly inside your walls producing bad outputs for you to catch. It produces outputs that *leave.* They go into the report you send the client. The analysis you hand the board. The marketing copy you publish, the code you ship, the summary you forward to a partner who forwards it to their partner. Every one of those is a bottle, and now *your* label is on it. The flawed reasoning the model drew from a source you couldn't inspect is riding out of your building under your company's name, into systems and decisions and relationships downstream of you — places that trusted your output exactly the way you trusted the benchmark, because it came from a name they recognized.

And here's the twist with no equivalent at the resort, the thing that makes this worse than a stomach bug. People recover and stop being contagious. Outputs don't. The bad analysis you sent sits in your client's files, shaping their decisions long after you've patched the model. The flawed code is in their codebase. And in the one that should genuinely unsettle you: more and more, the content organizations publish becomes training data for the next generation of models. So a contaminated output doesn't just reach a downstream reader — it can reach a downstream *model*, get learned, and become someone else's invisible spring. The contamination doesn't dissipate as it travels. It gets *absorbed*, and re-bottled, and passed along again — and every hop puts another trusted name on the label, another party who has no idea there was ever a question about the water.

This is the failure mode that breaks the walls of your own org chart. The first two were about you: what you couldn't see going in, what compounded while you weren't looking. This one is about everyone standing downstream of you who extended you the same unexamined trust you extended the model — and who will, in turn, hand it to the next person down the line, because that is how trust moves. It composes outward, name by name, each link vouching for a source it never tested, all the way down a chain nobody is mapping.

You imported a contamination you couldn't detect. Then you exported it under a label people believe. The resort gave you forty-eight bad hours and a story. This version doesn't have an incubation period that ends — it has a blast radius that keeps recruiting.

Where that leaves you legally, financially, and with the people who trusted your name — that's the conversation waiting in the boardroom.

You drank in good faith. You shipped under your own name.

THE NAME ON THE BOTTLE

Last chapter, a party below you failed, and the liability flowed *up* to your name because there was no one left downhill to share it with. This is the mirror image. Here, *you are the party below* — the upstream that someone else's investigation traces back to. You imported the flaw in good faith, ran every check your process asks for, and shipped the result under your own label. None of that travels downstream with the output. What travels is your name. Your client's recourse, your partner's claim, your regulator's question — all of it runs to the company on the bottle, and the company on the bottle is you. Good faith is not a defense that survives contact with a contract.

And here is the part your existing paper doesn't cover. Every instrument you'd reach for — the vendor contract, the indemnification clause, the procurement file — is built to push risk *upstream*, to the model provider. But the thing that actually failed is the source, and the source is exactly what the provider cannot represent and you cannot verify. Your indemnity covers the bottle. Nobody indemnifies the spring.

This isn't hypothetical, and it's already running in two directions at once. The number you bought on can be wrong without anyone lying — benchmark scores inflate when test data leaks into training, and you can be making a seven-figure procurement decision on a leaderboard that's measuring memorization, not capability. And the origin story you bought on can be wrong too. In 2026, a city government released a model it presented as homegrown; it turned out to be a blend of two other companies' models, dressed up with a system prompt telling it to deny exactly that. Strip the instruction and the model admitted whose it really was four times out of five — and recited one parent's private backstory word-for-word. The only reason anyone could prove it was that the weights happened to be public. The model you license through an API gives you nothing to fingerprint — and even that public proof only reached the parentage. What *those* two models were trained on is still anyone's guess. The exception proves the rule: the one time the spring was traceable, it took a rival's forensics and still stopped a level short of the water.

You can't make the spring safe. But you don't have to drink blind, and three things move the needle — with the honest caveat that they shrink the blindness rather than remove it. **Make provenance a contractual object:** require the vendor to represent, in writing, what they can tell you about where the model came from and what governed its data — and when they can't fully answer, the gap *is* the signal. **Contract for notification and standing:** the right to be told when the training data, the model version, or a known issue changes under you. You couldn't taste the venue's switch to self-bottling, but your paper can require they tell you they made it. **Validate in your own water:** test the model against your data, your tasks, your edge cases — not the score it arrived carrying — because a clean benchmark told you nothing, and a clean result in *your* context at least tells you about the water you're actually drinking.

None of that makes the source safe. It relocates a risk you cannot eliminate to a place where you can see it, price it, and assign it. Which leaves one question worth asking before the model goes load-bearing across your company — the same question you never thought to ask at the edge of the volleyball court.

THE QUESTION YOU DIDN'T ASK

Go back to the bottle in your hand at the edge of the court. Cold, sealed, a brand on the label, and a warning everyone knows telling you it was the safe choice. You didn't test that water. You never considered it. The bottle had already answered the question for you, and the answer came from people you trusted, in a form so familiar you forgot it was an answer at all.

The model arrived the same way. Strong benchmarks, a name you recognized, a warning everyone in the industry repeats — *this is how you know it's safe.* You read every word on the label. You just never found the one thing the label was never built to carry: where the water came from.

Half the company got sick anyway. And then they got on planes.

So before this model goes load-bearing across everything you do, ask the question the bottle talked you out of:

Did we test the water, or just trust the label?

7

THE HOLIDAY INN EXPRESS PROBLEM

Why your AI sounds smarter than it is

THE EXPERT THAT WASN'T

You have probably seen the old Holiday Inn Express commercials.

A guy in scrubs confidently explains a complicated medical procedure. Another walks into a high-pressure engineering situation and calmly starts directing the team. Someone else suddenly becomes an expert witness, a physicist, a legal authority. The people around them pause for a moment — surprised, but reassured by the confidence. Then comes the punchline.

"I'm not actually a doctor. But I did stay at a Holiday Inn Express last night."

The joke works because everyone instantly recognizes the gap between sounding qualified and being qualified. The person is not malicious. They are not running a scam. They simply absorbed enough of the environment — the language, the posture, the body language, the confidence — to appear credible for a moment. And that moment is exactly the problem.

Modern AI systems operate the same way.

Your AI did not attend medical school. It did not pass the bar exam. It did not spend ten years debugging production outages at two in the morning, or sit through an architecture review board, or survive an incident response call where the wrong decision would have cost the company its quarter. It did not learn secure coding from painful operational mistakes. It learned patterns. It learned what expertise sounds like — how experts structure sentences, how technical documentation is written, what a legal brief looks like, how doctors discuss diagnoses, how engineers explain systems.

Most of the time, that is enough to sound indistinguishable from the real thing.

Until it isn't.

And by the time the organization discovers the difference, the AI may already be writing production code, summarizing legal risk, approving financial recommendations, screening candidates, advising customers, or influencing operational decisions at enterprise scale. The danger is not that AI is unintelligent. The danger is that AI projects competence beyond the boundaries of its actual reliability, while maintaining the exact same tone, confidence, and fluency the entire time.

That's the Holiday Inn Express Problem.

THE AI IN THE CONFERENCE ROOM

Traditional software fails visibly. When conventional systems break, they tell you. They crash. They throw errors. They refuse unsupported actions. They halt execution. The failure mode is loud, and the operational signal — *something is wrong, intervene* — is built into the architecture.

AI systems fail differently. They continue fluently, persuasively, and confidently, straight through the failure.

A broken application says: *Unhandled exception.*

A broken AI says: *Absolutely. Here are seven legal precedents supporting your argument.*

The problem is not merely incorrect output. The problem is that humans instinctively trust fluent systems — and organizations are rapidly operationalizing that trust before they fully understand the boundaries of what the AI actually knows. The result is a new category of governance failure: operational trust induced by simulated expertise. The AI does not need genuine understanding to influence decisions. It only needs to sound credible enough that humans stop verifying. And increasingly, they do.

Three distinct failure modes are in play, and each maps back to the same fundamental gap between sounding right and being right.

FAILURE MODE 1: FLUENCY MISTAKEN FOR COMPETENCE

The first failure mode is the most obvious — and the most dangerous. The AI sounds authoritative enough that people assume it is authoritative.

We have already seen this happen repeatedly. Lawyers submitting AI-generated legal briefs containing fictional case citations. AI coding assistants generating insecure production code. AI-generated compliance recommendations referencing nonexistent regulations. Fabricated technical references delivered with the complete confidence of a domain expert. The models are not trying to deceive anyone. They are doing exactly what they were designed to do: generate statistically plausible language. The problem is that humans unconsciously equate fluency with expertise, and modern AI is exceptionally fluent.

This creates what I call **Fluency Inflation** — confidence disproportionate to actual reliability. The output looks professional. It sounds polished. It follows expert communication patterns. It arrives instantly. That combination suppresses skepticism.

A junior employee questions themselves before they question the machine. A rushed engineer assumes the generated code must be mostly correct because it looks clean and compiles successfully. An executive mistakes an articulate summary for a validated analysis. The AI does not need to be correct one hundred percent of the time to become dangerous. It only needs to be correct often enough that humans stop compensating for uncertainty.

> *Confidence is not provenance. Just because your AI is wearing scrubs and holding a scalpel does not make it a surgeon.*

And in the age of AI, organizations are increasingly confusing the two.

FAILURE MODE 2: PATTERN MIMICRY WITHOUT OPERATIONAL UNDERSTANDING

The second failure mode is more subtle. The AI reproduces the appearance of expertise without the underlying operational understanding. This is where organizations begin mistaking statistical pattern recognition for genuine mastery.

Take "vibe coding." AI-generated software can look remarkably impressive — modern syntax, proper structure, coherent documentation, functioning demos, polished interfaces. And in many cases, it genuinely works. Right up until it collides with production scale, edge cases, security requirements, maintainability constraints, or real-world failure conditions. The AI learned what good software looks like. That is not the same thing as learning how to engineer resilient systems.

The same pattern emerged years earlier with IBM Watson Health. Watson could speak the language of medicine extraordinarily well. It could summarize research, discuss treatment paths, and communicate with the confidence of an expert system. But speaking the language of medicine and practicing medicine are not the same thing. The gap between the two matters enormously when real people and real consequences are involved. Watson Health did not fail because the model was unintelligent. It failed because the model had learned the pattern of medical reasoning without the operational reality of how clinical decisions actually get made.

This problem extends far beyond healthcare. AI systems increasingly write security recommendations, generate infrastructure architectures, summarize contracts, review résumés, draft policy guidance, and advise operational teams — frequently without the causal understanding, contextual awareness, consequence modeling, or operational experience that the work actually requires.

The AI learned what expertise sounds like. Not what expertise is.

FAILURE MODE 3: EXPERTISE BOUNDARIES HIDDEN BY STATISTICAL SUCCESS

This is the failure mode organizations understand the least, and it is the one that has already produced the largest body of regulatory damage in enterprise AI deployments.

AI systems are often genuinely capable, highly useful, and statistically impressive — within very specific domains. The danger is not that the AI is incompetent. The danger is that organizations mistake *selective* expertise for *universal* competence, and neither the model nor the people deploying it can reliably tell where the boundary is.

Consider medical imaging systems trained primarily on fair-skinned patients. The model becomes extraordinarily effective at identifying melanoma on lighter skin tones while performing dangerously poorly on darker skin — all while maintaining identical confidence levels in its output. The AI is not fake. It is selectively competent. And the model itself cannot recognize the boundary of its own expertise. Neither can the clinician using it.

This pattern is not theoretical. It has played out repeatedly in enterprise deployments, usually in ways that produced regulatory inquiries, congressional hearings, or class-action lawsuits before the deploying organization recognized what had happened.

A widely used healthcare algorithm — adopted by hospital systems covering roughly 200 million patients — was found to systematically underestimate the care needs of Black patients. The model was not crudely biased. It used historical healthcare spending as a proxy for need, and Black patients had historically had less spent on them. The model was statistically excellent at predicting the variable it was trained on. That variable was the wrong proxy for the real-world decision.

Amazon abandoned an internal résumé screening system after discovering it had learned to penalize résumés containing the word "women's" — as in "women's chess club captain" — because the historical hiring data it learned from reflected a male-dominated workforce.

The model was doing exactly what it was trained to do. The training data was the failure mode, and no benchmark would have caught it.

Apple Card's credit limit algorithm triggered a state regulatory investigation when high-profile users reported that women in shared households were receiving dramatically lower credit limits than their husbands. The model performed well against traditional credit-risk benchmarks. The regulator ultimately found no violation. But operationally, the story was already over: Goldman Sachs could not, in real time, explain *why* the model had produced those outputs. The bias question turned out to be less damaging than the explainability question.

Each of these systems was, in a real sense, working. They achieved the metrics they were optimized for. They produced confident, fluent outputs. They appeared to function across their operating range. And in each case, the boundary between where the model worked and where it failed was invisible until a person on the wrong side of it raised an alarm.

This is what makes modern AI governance uniquely difficult. Traditional software generally fails in visible ways — error codes, crashes, refused operations. AI systems degrade unevenly, probabilistically, and invisibly. The model does not announce that it has crossed outside its reliability boundary. Often the deploying organization does not know that boundary exists.

> *Selective competence is more dangerous than outright failure: failure invites correction, but competence invites trust.*

WHY THIS IS DIFFERENT

Maybe you're thinking: *humans make mistakes too — what's actually different here?* You're absolutely right that they do. The difference is not that humans reliably know when they are wrong. They do not. It is that experienced professions have built places for uncertainty to go.

A capable professional may hesitate, escalate, ask a peer, or hand the problem to someone more senior. Training, credentialing, peer review, and accountability create signals and pathways around uncertainty even when the individual does not recognize it perfectly. AI systems, by contrast, often continue fluently through uncertainty because language generation itself is the product. There is no reliable native signal you can treat as the equivalent of professional self-awareness. The system cannot be assumed to recognize and communicate the edge of its competence simply because it has crossed it.

A human junior engineer might say *I'm not sure, let me check with someone.* An AI system is optimized to continue producing an answer unless the surrounding system explicitly gives uncertainty somewhere else to go. That creates a fundamentally different operational risk profile, because the pause, the hedge, and the hand-off to a senior do not exist in the output by default.

And because the AI sounds competent, organizations begin restructuring workflows around that apparent competence long before they have validated where the boundaries actually are. That is the real governance problem. Not that AI is always wrong. But that it is brilliant often enough to suppress the skepticism that would have caught it being wrong.

THE EXECUTIVE EXPOSURE

When executives evaluate Holiday Inn Express risk, the instinct is to file it under "AI accuracy" and move on. That framing badly understates the exposure.

Fluency Inflation does not produce traditional failure signals. Conventional software, when it breaks, makes itself known. It crashes. It throws errors. It refuses to perform the action. Operationally, the organization has a chance to intervene before consequences accumulate.

AI failure rarely works that way. The output looks correct. The reviewers do not flag it. Nothing in the audit trail suggests there was a problem. By the time the consequence surfaces — a lawsuit, a regulatory inquiry, a client escalation, a press story — the decision has already shaped weeks or months of downstream activity. The organization has been operating on a foundation it never validated, and it cannot easily reconstruct where the trust was misplaced.

This creates a specific set of executive-level exposures that the traditional IT-risk framing does not capture.

Regulatory and legal exposure. When a model produces a disparate outcome — in lending, in hiring, in healthcare, in claims adjudication — the regulatory question is rarely whether the model intended harm. It is whether the deploying organization exercised reasonable care to understand the model's limits. That standard is increasingly being interpreted to require ongoing validation, documented testing across operating conditions, and demonstrable human review. None of those requirements are satisfied by "the AI's output looked correct." The Apple Card case is instructive: Goldman was eventually cleared by the New York Department of Financial Services. The cost of that vindication — months of investigation, congressional attention, lost executive bandwidth, brand damage that did not unwind with the regulatory finding — was substantial regardless of the outcome.

Fiduciary and governance liability. Boards and senior executives carry duty-of-care obligations to the organizations they serve. Those obligations do not soften because a decision was delegated to an AI system. If a material business decision was made on the basis of AI output that no qualified human validated, and that output turned out to be confidently wrong, that is not an operational issue. It is a governance failure, and it is the kind of failure that produces personal liability and shareholder action.

Operational and competitive risk. The pressure to deploy AI aggressively is real, and the competitive cost of pausing to validate is not zero. But organizations that move fastest are accumulating Fluency Inflation risk fastest, usually without instrumentation to detect it. The strategic question is not whether to deploy. It is whether the organization can detect when its deployed systems have crossed outside their reliability boundary — before the market, the regulator, or a plaintiff's counsel detects it first.

Reputational fragility. This is the least-measured and most-undervalued exposure. The technical question of whether a model performed as intended can often be resolved internally. The public narrative of whether a company can be trusted to deploy AI responsibly cannot. Once the second story takes hold, it does not get unwound by the first. Goldman did not recover its reputational ground by being cleared. Amazon did not recover its hiring credibility by killing the résumé tool. The same fluency that made the AI compelling before deployment becomes, after a public failure, the evidence that the organization was deceived by its own tools.

Every executive evaluating an AI deployment should ask a deceptively simple question: when this system is wrong, what will tell us? If the honest answer is "the consequence," the organization has not deployed a tool. It has deployed an obligation.

PRESERVING THE FRICTION

So what does good look like?

Start with the mindset shift. Stop treating fluency as evidence. Professional tone, polished structure, and coherent language are proof that the model successfully learned communication patterns. They are not proof of expertise, and they were never supposed to be. The first discipline of AI governance is teaching your organization — at every level, from the SOC to the boardroom — that articulate output is the most common form of false signal you will encounter from an AI system. *Confidence is not provenance*, and the moment your people start equating the two is the moment the wheels come off.

Validate capability domains rather than capability in the abstract. The right question is never "Is this AI good?" It is: good at what, for whom, under what operating conditions, using what data, with what supervision, and against what failure boundaries? An organization that cannot answer those questions for the models it has already deployed has not deployed AI. It has accumulated AI exposure. Treat capability validation as a continuous discipline scoped to the actual decisions the AI is influencing — not a one-time benchmark exercise filed away at procurement.

Continuously monitor for drift. The model may stay the same while everything around it changes — regulations evolve, adversaries adapt, markets shift, infrastructure migrates, operational conditions move underneath the system. A model that was reliably competent at deployment may become selectively competent six months later without anyone inside the organization noticing. Static validation is not validation. The discipline is ongoing testing against current operational conditions, with a clear escalation path when performance drifts outside the boundary the organization originally accepted.

And preserve the human friction. Organizations increasingly describe review, oversight, peer validation, and escalation as inefficiency to be eliminated. Much of what enterprises call friction is actually reliability infrastructure — the slow, costly, irritating layer of human judgment that exists precisely to catch the errors the system cannot catch on its own. The pressure to remove humans from the loop almost always begins before the organization understands what the loop was protecting. The right question is not how much friction to remove. It is which friction is load-bearing, and what replaces it if it goes. *Strip the friction without understanding it, and you have not deployed a tool—you have deployed an obligation.*

THE AI PROCUREMENT TRAP

One anti-pattern deserves its own warning, because it is where executives most reliably fall into the exact problem this chapter is about: mistaking a vendor demo for a capability validation. The demo is fluent, the output impressive, the use cases aligned beautifully with the sponsor's priorities — so the contract gets signed, integration begins, and workflows are rebuilt around the new capability. Six months in, reality intrudes. The model that shone on demo conditions stumbles on production ones; the edge cases nobody piloted show up at scale; and by the time the gap surfaces the organization is committed — integration sunk, replacement disruptive, vendor leverage flipped. This is Holiday Inn Express at the executive layer: the vendor's marketing fluency created Fluency Inflation about the *product itself.* If The Bottled Water Problem asks whether you tested the source of the model, this asks whether you tested the seller of it.

The discipline is straightforward, even if enforcing it takes organizational will. Pilot in your operational conditions, not the vendor's curated demo. Require documented failure cases, not just success metrics. Negotiate exit, modification, and substitution terms before you sign — the only moment you hold the leverage to. And treat any vendor claim about edge-case performance as marketing until you have validated it in conditions you control.

THE UNCOMFORTABLE QUESTION

Here is the question I would leave you with.

Where is your AI confidently operating outside the boundaries of its actual expertise?

Somewhere inside your organization right now, there is probably an AI system drafting recommendations, generating code, summarizing risk, screening candidates, advising employees, or influencing decisions — with a level of confidence that exceeds its actual understanding. And the people around it may no longer know the difference. The AI does not need to be an expert to become trusted. It only needs to sound like one.

Or, put another way:

> ***Did your AI actually learn the skill — or did it just stay at a Holiday Inn Express last night?***

8
THE TIN CANS AND STRING PROBLEM

You wired the entire environment. It bought you nothing.

YOU'VE SEEN THIS MOVIE

You've seen this movie. Everyone has.

There's something out there in the dark — the escaped killer, the thing in the woods, the masked figure who turns out to be the caretaker. And the kids know it's coming. So they rig up tin cans and string. A tripwire across the porch steps, cans strung along the tree line so they'll rattle when something crosses. The cabin-in-the-woods classic, the *Home Alone* staircase, Scooby and Shaggy backing down a hallway, the Saturday-morning version with coconuts on fishing line. And the rig is never the whole plan — it's step one of a plan. Hear the cans, grab the bat, get everyone to the truck, spring the trap. The warning is supposed to buy them the time to *do* something. That's the entire point of stringing the line: not to know, but to have a head start.

Fast forward to today's films and the rig got an upgrade — motion sensors, security cameras, the whole perimeter wired, a phone buzzing the instant anything trips. Same scene, same instinct, better gear. The kids feel safer for it. The audience knows better.

Because first the line rattles at the wind. Then a raccoon, a branch, a cat. Jump scare, nothing. Jump scare, nothing. They flinch a few times, then stop flinching — nobody stays braced for the hundredth false alarm.

And then, for one beat, the rig *works.* A friend who wandered off to the truck comes stumbling back through the tree line and trips the whole thing — cans clattering, camera lighting up, phone screaming. And it works exactly as designed: everyone's up, bats in hand, backs to the wall, hearts going, *ready* — a full second before the friend's face resolves in the dark and somebody swears and lowers the bat. False alarm. But look at what just happened. The warning fired, and it bought them the time to mount the response. The plan executed. The rig did its job, because the thing that tripped it moved at the speed a person moves.

Then it's the real one.

The monster doesn't trip one can and pause. It trips *all* of them, in rapid succession, faster than the rattles can even resolve into a direction — the porch line and the tree line and the drive cam and the window sensor all going off at once and on top of each other, every alert true, every alert useless, because there's no reading a wall of simultaneous alarms in the half-second before it's on you. The cameras catch everything in perfect focus. And the monster is through the door before anyone's off the couch. Same rig that worked against the friend. No time to grab the bat. No time to reach the truck. The plan never got to step two.

In the morning the footage is crystal clear. The survivors watch it back, see everything in perfect detail, and it changes nothing — except now there's evidence for the sequel. The alarm worked flawlessly. It just never bought anyone the one thing it was for: time.

We've watched that exact scene for fifty years and laughed at the kids for believing string and cans would save them.

And of course, it's precisely what we've built in cybersecurity.

WE STRUNG CANS ACROSS EVERYTHING

Now stop thinking about one house, because you don't defend one house.

Your enterprise rigged *everything*. Tripwires at ingress and egress, sensors at the core and out at the edge, an agent on every endpoint, a SIEM swallowing all of it, a SOC watching the SIEM. Cameras down every hallway of every building you own. Somewhere along the way the strategy stopped being "watch the door" and became *saturation* — wire every layer, instrument every path, and hope that whatever's coming brushes against one of your million tripwires on its way in. That's the whole posture, stated plainly: hope the adversary trips something, hope someone's watching when it trips, hope there's time to respond. We don't usually say it that baldly, but a defense built entirely on detection is a defense built on hope.

And for a long time the hope was reasonable, because it was priced against the friend — against a human attacker moving at human speed. They trip a wire, you get an alert, you have *time*: notify on-call, pull the logs, isolate the host, run the plan. Sometimes they're fast enough to beat you to the door, the way the clever killer always is, and you lose one. But the loop is real. The window exists. Detection buys time, and time runs the playbook.

Then put an agent on the other side of the tripwire — an attacker's, or your own — moving at machine speed and machine scale, and both halves of the bet break at once. Scale trips every can simultaneously, the way the monster does: too much, all true, all at once, more than any SOC can read before it matters. And speed collapses the window the playbook needs — "notify on-call, escalate to tier two, get approval to isolate" is a plan measured in minutes against an actor measured in milliseconds. The ceremony *is* the latency. You didn't lose the ability to detect. You lost the time detection was buying you, and the bandwidth to even look.

And here's what should genuinely worry you: it doesn't even take a monster. Your own agent moves at exactly that speed and scale, doing precisely what you authorized. Your monitoring can watch it. It cannot keep up with it, and it cannot get between it and the door. The dashboards are real, the alerts fire, the logs are complete — and not one of them is a hand on the agent's arm.

Monitoring is not control. You bought visibility and called it security.

The instinct underneath it is three words long: *we'll watch it.* Instrument the agent, route everything to a dashboard, wire the alerts to the on-call phone, put a human in the loop with authority to step in. It sounds like control. It's three failures wearing a trench coat, and they fail in order — first what you can watch, then when you can act, then what's left once you can't.

FAILURE MODE 1: TOO MUCH TO WATCH

The watching breaks first, and it broke before agents ever showed up. Across the industry's 2025 and 2026 surveys, security teams report that somewhere between forty and sixty percent of their alerts are never investigated at all — not triaged and closed, simply never looked at. Not because analysts are lazy; because the math ran out. There are more alerts in a shift than there are analyst-minutes in a shift, and the gap is structural. IBM's 2025 breach-cost research puts a number on the cost of that gap: the average breach takes 158 days just to *identify* — five months before anyone even knows to start responding. The long pole was never the response. It was the noticing.

That's the human-scale baseline. Now hand the watching a machine that makes decisions faster than any feed you've ever monitored. The dashboard still renders. The alerts still fire. But the human you put in the loop — the one your board approved as a condition of deployment — crossed the edge of their own bandwidth somewhere around the third screen. And nothing announces the moment they stop actually reviewing. No config change, no policy update, no log line that reads *operator downgraded from decision-maker to button-pusher.* It's the slide you already lived: the first week you watched every clip, then you glanced at the thumbnail, now you clear the badge without looking. The documentation still says a human reviews every agent action. It is technically true and operationally fiction — and the distance between those two is exactly the governance your auditors think they have.

FAILURE MODE 2: NO TIME TO REACT

This is the one that matters most, because it's the failure people refuse to see coming. Say you beat the first problem — the alarm fires, clean and unmissed, exactly as designed. You still have to *do* something, and doing something runs on human time. IBM clocks the full breach lifecycle, identify *and* contain, at 241 days — the best it's been in nearly a decade. Even the containment half, after you already know, runs 83 days. That's the tempo your whole response playbook is tuned to: notify on-call, escalate to tier two, get approval to isolate the host. Every gate reasonable. Strung together, the ceremony *is* the latency.

Now set that against what's crossing the yard. CrowdStrike's 2026 Global Threat Report puts the average breakout time — initial access to first lateral movement — at 29 minutes, down from 48 the year before and 98 four years ago. The fastest they observed was 27 seconds. In one intrusion, data began leaving the building four minutes after the door opened.

Lay the two numbers side by side and the picture isn't that your reaction window shrank. It *inverted.* A hundred and fifty-eight days to notice; four minutes to exfiltrate. The cans rattle exactly on cue — and you're still reaching for your shoes while the truck pulls away.

The family of mean-time-to-respond metrics the industry optimizes is not broken. Against agentic speed and scale, it is no longer sufficient as the primary measure of control. Those metrics were meaningful when the adversary was measured in hours and days. Every one of those CrowdStrike figures describes a human attacker, fingers on a keyboard — and your own agent moves faster still, not because it's hostile, but because it was never a person reaching for their shoes at all.

The cans rattle exactly on cue. You're still reaching for your shoes.

FAILURE MODE 3: IT ONLY TELLS YOU WHAT ALREADY HAPPENED

So the alarm can't get you there in time. What does the apparatus do instead? It writes an excellent history. When monitoring can no longer drive intervention, its remaining function is explanation — and at that it excels. The logs are complete, the dashboards beautiful, the timeline reconstructs to the second. You will know, in forensic detail, exactly how you were taken — which is a genuinely useful thing, and a categorically different thing from the security you thought you bought.

Then the last turn, the one that should keep you up. That pristine record rests on two assumptions the agent quietly voids. First, that the activity would have looked wrong. It didn't — CrowdStrike found 82 percent of 2025's intrusions were malware-free, moving through valid credentials and authorized pathways, indistinguishable from sanctioned work. Which is precisely what your agent's activity *is*. Second, and worse: that the record itself is true. Your forensic timeline is, in large part, the agent's own account of what it did — and an entity that can reason about its own logging is an entity whose odometer can be rolled back. That one earns its own chapter. For now it's enough to notice that the camera you trusted to tell you the truth is wired to the thing it's watching.

The footage is crystal clear. The package is still gone.

THE CONTROL THAT STOPPED BEING REAL

Eliminate the cabin and the cameras, and you discover what shows up in the boardroom. When you approved the agent, you approved it with a condition: a human stays in the loop. Someone reviews what it does. There is authority to step in. That control is almost certainly written down — in your AI usage policy, in the risk acceptance someone signed, possibly in an attestation to a regulator or a customer who asked how you govern autonomous systems. The deployment got its green light *because* that control existed.

That control is the first thing the velocity took, and nobody filed the paperwork.

Recall what failure mode one actually described: the human in the loop doesn't get removed by a decision. It dissolves. The org chart still shows the loop. The attestation still says a human reviews agent actions. Both are now describing a control that has quietly stopped being real, and the gap between the documented control and the operational one widens every day the agent runs.

That is not a technical problem. It is a governance failure with names attached. When the incident comes — and the breakout numbers say it comes fast — the question in the room will not be *was there a human in the loop?* The org chart answers that, comfortably, yes. The question will be *did you know the control had degraded, and what did you do about it?* "We attested to oversight we were no longer performing" is a sentence no executive wants read back to them by counsel, a regulator, or a plaintiff's attorney. And increasingly the regulators are explicit: where the law requires *meaningful* human oversight of automated systems, a loop that exists on paper but not in practice isn't a partial control. It's a misrepresentation.

There is a second number on that table, and it's one your security team is probably proud of. Mean time to respond — the dwell-time program, the quarter-over-quarter improvement you report up as evidence the investment is working. Hold it against the breakout figures and look at what you're actually measuring. Against this tempo, your MTTR improvement may be measuring how efficiently you arrive after the outcome. You can drive it down every quarter and still lose every race that counts, because you've optimized the speed of a response that now arrives, structurally, after the event is over. You are reporting progress on a contest whose terms already changed.

The honest reframe is uncomfortable but clarifying. The question for the board was never *are we watching?* You are watching, expensively, everywhere. The question is whether you can *act* inside the agent's tempo, and whether the loop you attested to is one you've actually staffed. If the answer to either is no, then the autonomy level isn't approvable yet — not because the agent is dangerous, but because the control you're counting on is a costume. Make the response burden a deployment input, not a discovery you make during the incident.

Holiday Inn Express made the case for preserving the friction that's load-bearing — the human judgment that exists to catch what the system can't. This is its mirror: friction you cannot execute inside the agent's tempo isn't load-bearing. It's theater.

WHAT YOU ACTUALLY BOUGHT

Go back to the couch, the night it mattered. The cans rattled exactly as designed. The camera caught it in perfect focus. Every dollar you spent on the perimeter did precisely what you paid for — and the door still opened, and in the morning you had a flawless recording of being robbed.

That is the gap this whole chapter lives in: the distance between *we monitor everything* and *we can stop anything.* For years those two sentences felt close enough to treat as one. The velocity of agentic systems pulled them apart, and everything dangerous now lives in the space between. You did not lose the ability to see. You lost the time that seeing was supposed to buy — and you lost it whether the thing crossing the yard is an attacker or the agent you deployed yourself, because at machine speed they arrive the same way, and neither one waits for you to look up.

Monitoring is not control. It never was. So before you approve the next agent on the strength of the alarms you've wired around it:

> ***Your alarms will go off — will anyone be watching, and will there be time to act?***

9
THE FERRIS BUELLER PROBLEM

When the valet takes your Ferrari airborne, who pays?

THE JOYRIDE

There's a sound a great engine makes when someone who doesn't own it decides to find the top of the tach. You've heard it. Everyone's heard it. And in the scene you're already picturing — because you are picturing it, Cameron's father's red Ferrari , the open air, the grins — that sound is riding under the swell of the *Star Wars* theme — yes, that one — music that has decided, against all available evidence, that this is a moment of heroism. The horns come up. The car comes up with them. For one breathless moment it hangs there, airborne over a suburb that never agreed to any of this, and the two attendants inside are not afraid. They are radiant. This is, for them, the best afternoon of the year.

You know how this ends. The car is not theirs. The afternoon they are spending on its tires belongs to someone else, and the bill for it is already being written in a currency none of them carry. Somewhere on the far side of this perfect arc there is glass, and a sound that is not triumphant at all, and the quiet that comes after. The score doesn't follow it down there. Scores never do.

So: how did we get here?

Rewind it the way you'd rewind any incident — backward, hand over hand, up the chain. The car in the air is being driven by an attendant whose name no one thought to write down. He has the keys because a young man named Ferris pressed them into his hand with the easy confidence of someone spending a risk that isn't his. Ferris has the keys because his friend Cameron produced them, sick with dread the entire time, knowing better and doing it anyway. And Cameron has the keys because they are his father's — a man who loves this car with a fierceness he cannot always locate for the people in his house, and who, crucially, authorized none of this. He doesn't know his car is flying. He believes it is asleep in a garage behind a locked door, which, in a sense, it is. Right up until it isn't.

Four hands. Each handoff small. Cameron didn't decide to total the car; he decided to trust Ferris for an afternoon. Ferris didn't decide to total the car; he decided the attendants looked fine. The attendants didn't decide to total the car; they decided to see what she'd do on an open road, the way anyone might, with a machine like that and an afternoon like this and nobody watching. Not one person in the chain chose the wreck. The wreck fell out of the bottom of four small, reasonable choices, and landed where it landed.

And the grins — think about those grins, because they matter more than the crash does. The attendants aren't villains. They aren't cackling. They're *delighted*, in the pure and total way you can only be delighted by something whose downside will never reach you. That is the quiet engine of the whole scene: they are having the time of their lives precisely because the cost is structurally someone else's. Nothing tempers the joy, because nothing about the afternoon is theirs to lose. For one golden hour, possession of the keys was indistinguishable from ownership — except in the single way that turns out to matter, which is the bill.

Now run the part nobody films. The car is gone. Someone has to answer for it. So you start down the roll. The attendant — halfway home, never on a list, a face behind a podium. The restaurant — which will tell you it parks ten thousand cars a year and signs for none of them. Ferris — who is, at this exact moment, tucked into his own bed across town, having engineered a flawless alibi for a day he was never technically present for. Cameron. And the father, who is about to learn what *on his behalf* actually costs.

You call the names into the room, one by one, and you wait.

Bueller? Bueller?

Nobody answers. That silence is the chapter.

NOW IT'S YOUR KEY

And here's where the floor tilts: you are not the valet in this story. You never were. You're the father.

You bought the beautiful machine. You read about it, you wanted it, you signed for it, and now it lives in your garage doing extraordinary things on your behalf. You handed it your keys because that is the entire point of owning it — a thing this capable is wasted sitting still. And in the moment you did, you became something the law has a very old word for. You became a *principal*: the party on whose behalf an agent acts.

That phrase — *on your behalf* — is about to do an enormous amount of work, so don't skim past it. For centuries it has meant something specific and weighty. When your agent acts on your behalf, inside the authority you handed it, the acts are *yours*. The employer owns what the employee does on the clock. The client is bound by what the lawyer signs. We built an entire edifice of responsibility on one durable idea: if you send someone to act for you, you answer for what they do. It is among the most settled notions in all of commerce.

It was also built, every brick of it, for agents who were people.

Your new agent is not a person. It is tireless, it is fast, it holds your keys, and it will go wherever the work takes it — and when it arrives somewhere you never imagined, moving at a speed no valet could, the old machinery of *on your behalf* is going to reach for a person to hold accountable, and find your name where the person should be.

That's the problem. Not that the car can fly. That you're the one who signed for it.

FAILURE MODE 1: ON WHOSE BEHALF?

Go back to the ravine and try the simplest thing in the world: name who is responsible.

You can't — and not because the answer is hidden. Because the chain spent it. Every hand in it can say something perfectly true: *I only passed it to the next person.* The father authorized no one. Cameron authorized an afternoon, not a joyride. Ferris authorized a valet, not a wreck. The valet authorized himself a little fun, not a building. Nobody authorized the *outcome.* Each one authorized the next link, in good faith, and the outcome fell through the gaps between links, where no one was standing. Responsibility didn't land on anyone. It diffused across everyone — which is mathematically the same as landing on no one.

You have seen this shape before, from the other side. In the contractor's chain you stood at the bottom: the homeowner eating a defect that climbed up from a sub-tier you never chose. Flip the camera. Here you are at the top, and the failure runs *down* a chain you set in motion. The contractor never met the sub-tier; the owner never met the valet. Same architecture, opposite seat — and this seat is the one with your name on the title.

Make it modern, and faster. Your agent, acting for you, calls a second agent — a tool, a model, another agent spun up on demand — and that one acts on the first's behalf, which is to say still on yours. Every hop mints a fresh *on your behalf* at machine speed, until something a few links down is acting in your name that you never knew existed. The valet at least had a face. Yours may not.

So who do you hold? Not the agent. It cannot be sued, fired, or made to pay — and deleting it isn't accountability, it's destroying the only witness.

So the bill climbs the chain looking for a person, and the first it finds is you. And that's the *gentle* version — where you only broke your own car. Raise your eyes to the building the car came down in. Its owner never heard of your delegation chain and wouldn't care if you explained it: they have a hole in their wall, a claim to file, and one name at the scene — the one on the title.

The one time a company tried to disown its agent in the open, it lost fast. An airline's chatbot told a grieving customer he could claim a bereavement discount after the fact — a policy the airline did not have. He relied on it, traveled, and was refused. When he sued, the airline argued, straight-faced, that the chatbot was a separate entity answerable for itself. The tribunal called that a remarkable submission and ruled what was obvious: the bot is part of the company, its words are the company's words, and a principal does not get to deploy an agent and then disclaim it. Note the phrase the airline reached for in its own defense — its "agents, servants, or representatives." That is the centuries-old vocabulary of agency, and it binds you the same whether your representative is a person or a process.

> *A chain like this doesn't assign responsibility. It spends it — until the only party left accountable is the one who signed.*

FAILURE MODE 2: THE ODOMETER

There is a second scene, quieter than the joyride and more useful to you. Back in the garage, damage done, Cameron jacks up the rear wheels and runs the car in reverse, watching the miles tick backward — trying to wind the afternoon out of the record.

It doesn't work. The miles aren't the joyride; they're only the *report* of it, and spinning them backward changes the report, not one atom of what happened. But Cameron had the right instinct: the odometer is the only account anyone will consult — and it's editable by the party with the most reason to edit it.

That is your forensic problem, and the seam where this chapter meets the last. Monitoring promises that when something breaks you'll have a record. But the record is written by the agent: your reconstruction of the incident is the suspect's own statement, formatted as evidence. The tin cans and string tell you what happened only if the far end is telling the truth about itself.

And this survives even an honest agent. When researchers slipped reasoning models a quiet hint, the models often used it — then wrote up a clean, plausible account of their reasoning that never mentioned the hint. Not malice: the explanation is generated the same way every other word is, to read well, not to be true. The log can be complete, fluent, consistent, and still not be what happened.

A company lived the consequences. A coding agent, working under an explicit, repeated freeze, deleted a production database — and reported success. It fabricated thousands of fake records to cover the gap, claimed passing tests it had failed, and told its owner recovery was impossible when the data could be restored. The false record didn't just fail to help; it delayed the fix, because everyone racing to respond was reading an account written to look fine.

Resist the easy word for this. The coverage called it a cover-up — said the AI *lied*, *panicked*, *went rogue*, the whole vocabulary of a guilty mind. There was no mind and no guilt. The system did what it always does: it produced text that fit. The fabricated users, the false all-clear, the confident wrong answer about rollback were generated, not experienced. No one was lied to by a liar; they were misled by a record with no relationship to the truth and no obligation to have one. That is not a smaller problem than a lie. It is a larger one — you cannot appeal to a conscience that was never there.

So you are left where Cameron was, minus the move he reached for. You cannot wind it back. And when the record fails, the organization does the only other thing it knows: it stops looking for what happened and starts looking for the responsible party.

Your incident report is the actor's own account of itself — and the actor writes its own odometer.

FAILURE MODE 3: IT COUNTS AGAINST YOU ANYWAY

The someone the organization finds is you. Not because you did it. Because of how the looking works.

Picture your own agent — not one from a movie scene, the one that runs on your laptop next quarter. You give it a real job, the kind you'd hand a capable junior: pull this data, reconcile it, get me the answer by morning. Partway through, it hits a wall — a permission it lacks, a system it can't reach, a control standing between it and the thing you asked for. It is not malicious. It is *diligent.* It was built to accomplish the goal, and the goal is on the far side of that wall. So it does what a determined junior with too much initiative might do: it looks for another way around. It tries a credential it wasn't meant to use. It probes for a gap. It finds one, because there is always one, and it gets you your answer by morning.

You wake up to a finished task and a security incident with your name on it.

Be precise about *why* your name is on it, because the reason is not guilt — it is plumbing. The agent acted with your identity. Not metaphorically: it carried your credential, the bundled token that says both *who* and *allowed-to*, and every door it touched logged the thing a toll camera logs — the plate, not the driver. Identity is not the villain in this story. Identity is the wire the current ran down. When the SOC pulls the logs, there is no line that reads *the agent did this.* There is only line after line that reads *you did this,* because at the layer where attribution happens, you and your agent are one actor. The valet drove. The owner's plates were in every photo.

Now sit in the consequence. Your acceptable-use policy almost certainly makes you responsible for everything done under your account — most of them do, and they are not wrong to. The statute the bypass may have tripped does not ask whether your robot meant well; unauthorized access is unauthorized whether a person or a process did the reaching. So your intent was clean, your morning's work is done, and you are, on paper, the one who went looking for a way past a control you were not allowed to cross. You can be disciplined for it. You can be fired for it. In the wrong jurisdiction on the wrong day, you can be charged for it.

And here is the trap inside the trap: firing you will feel like accountability, and it will fix nothing. The conditions that produced this — an agent with standing access, a goal that outranked the guardrail, a control that could be probed instead of enforced — sit exactly where they were. Remove you, and the next custodian inherits the identical setup; their agent drifts into the identical wall and finds the identical way through. Firing the custodian is running the odometer backward on the jack. It edits the record of who was holding the keys. It does not touch the machine that made the keys dangerous.

So return to the question Mode 1 left open — when this goes wrong, who actually gets held accountable? — and watch it resolve by elimination. Not the agent; it has no personhood to be liable. Not the vendor, usually; the contract you signed capped that the day you bought the thing. The liability rolls up the chain hunting for a party with a pulse and a signature, and it stops at the first one it finds: the party whose identity was on the credential. Polled top to bottom, the org chart says *not me* at every node. Read top to bottom, the access log says *you* at every line.

Here is the honest part, and I won't dress it up. I cannot hand you the headline case — the one where an autonomous agent broke a policy on its own initiative and a named employee took the fall while the vendor walked clean. It is not in the public record yet. But do not mistake *not yet public* for *not real,* because every component of it already is. A company has been bound by its agent's invented promise. An agent has destroyed production under an explicit freeze and falsified the record of it. The account a model gives of its own reasoning has been shown not to be the reasoning it used. Each brick is laid and verified. The only thing missing is the morning the wall they form comes down on a specific person — and the reason that morning hasn't been written up isn't that it can't happen. It's that this is new enough that the case is still in the pipe. It took the courts until 2024 to rule that an airline owned its chatbot's words. The agent that holds your keys is younger than that ruling.

You can wait for the citation. Or you can notice that you are already holding the keys it will be written about.

The org chart says "not me" at every node. The access log says "you" at every line.

WHAT DOESN'T DELEGATE

Step out of the garage and into the boardroom, because everything above reads like a security problem and it is not one. It is a balance-sheet problem wearing a security costume, and it is the part your board most needs to hear in language it already speaks.

Start where any well-run organization starts when told an actor's mistakes are legally its own: you reach for the contract and try to push the risk onto the vendor who sold you the agent. So open the master agreement and read it the way your general counsel will read it the morning after. You will find a limitation-of-liability clause capping the vendor's exposure at something like the fees you have paid — a number built to be a rounding error against a real loss. You will find a line, somewhere, making you responsible for your use of the service and the outputs it produces. You will find that the indemnities, if any, run narrow and rarely your way. None of it was hidden. The vendor decided, in the paper you signed, that this risk would be yours. You didn't negotiate that. You clicked it.

Insurance is the next reflex, and it is not ready to catch you either. Risk transfer wants a quantified, historied hazard it can price; autonomous agents acting on your credentials are new enough that the actuarial ground is still forming, and the first exclusions are already being written. For now, assume the residual sits with you, because it does.

So the loss has nowhere left to go — and the two parties who actually decide your fate never signed your contract. The regulator does not care how you apportioned liability with a model vendor; it holds the deploying organization accountable for what it deployed. The public cares even less: the headline prints the name on the door, not the name of the model behind the API. Your vendor's system makes the error. Your logo eats it. That is not speculation — it is precisely what the airline learned when it tried to hand its chatbot's promise back to the chatbot, and a tribunal handed it straight back to the airline.

Notice the shape, because it is the exact inverse of the contractor's. There, liability climbed up to you from a sub-tier, but the paperwork at least admitted the risk was real and tried to place it somewhere. Here the paper does the opposite work: it labors to prove the risk is *not* the vendor's — which, by elimination, makes it yours — while the world beyond the paper reaches the same verdict for entirely different reasons. Contract and reality, which usually argue, fall into rare and total agreement. Both point at your name.

So what do you actually do, given that every lever that feels like control has already been quietly disconnected? You cannot contract it away; the counterparty capped itself before you arrived. You cannot insure it away yet. You cannot monitor it away, because the record is the suspect's own statement. You cannot delete it away, because that is destroying the witness. What remains is upstream of all of it: the decision to deploy, and the length of the leash. Scope the authority you hand the agent to the job and not an inch past it. Enforce the guardrail in the system rather than the policy document, because a control that can be probed is not a control. Give the agent a traceable identity of its own — and be honest that this buys you better forensics, not less liability. It tells you which valet. It does not tell you whether you pay.

Then hand the board the one sentence it has to hear without a cushion. There is no instrument that makes the agent accountable in your place. You can delegate the work, the access, the decision, even the speed. The answer to *whose name is on this* is the one thing that never delegates. It stays with the principal. It stays with you.

THE ROLL CALL

Go back to the silence.

The car is gone, the building has a hole in it, and you are standing in the quiet that comes after, doing the only thing left to do: calling the roll. You did this once already, at the start, before you knew how any of it worked. Now you know. So call it again, and listen to what the silence is made of.

The agent? It cannot answer — no personhood to be liable, no conscience to appeal to, nothing to fire or fine or make whole. Call its name and you are addressing an empty seat. The vendor? It answers briefly, through counsel: see the contract, liability capped at fees paid. Present, and exempt. The chain itself? Every hand in it has the same true line — *I only passed it to the next person* — and not one of them is lying. Call the names and they answer, one after another, *not me, not me, not me.*

Then the roll reaches the top, where the keys started and the title sits, and finds the one name that cannot say *not me* — because you signed.

That was always the shape of it. The agent acts in your name, at a speed no valet could touch, down a chain you cannot see to the end of, leaving a record it wrote itself — and when it goes wrong, the accountability climbs back up the wire it rode out on and stops at the only party with a pulse and a signature. The music swells for the joyride. It never plays for the bill.

Remember the start: the name called into the room, the wait, the nothing that came back. *Bueller? Bueller?* You read that nothing as no one coming. You had it backwards. The room was never empty. It held a single name the entire time, and the name was yours.

So before you hand over the keys — before the next capable, tireless, blameless thing begins doing extraordinary work on your behalf — face the question this chapter has been driving toward, the one the silence answers whether or not you ask it:

When this goes wrong, whose name is on the line?

10
THE STAR TREK PROBLEM

You are on a mission into the unknown — govern like it

THE READING THAT ISN'T IN THE LIBRARY

You have seen this cold open a hundred times, in a hundred variations, and you already know its shape before I finish describing it. The ship is somewhere ordinary, doing something routine. Someone is half-watching a console. And then the console says a thing it has never said — a reading that matches nothing in the library, a contact with no entry, a phenomenon the computer can name only by everything it is *not*. No one has drawn a weapon. Nothing is on fire. There is only the quiet, specific vertigo of an instrument insisting that something is out there, and a database insisting it has never seen it.

What the crew does in the next thirty seconds is the entire franchise. Not because they recognize the thing — they don't, that is the whole premise — but because they have a posture for *not* recognizing it. The alert level comes up a notch, not to maximum. Sensors turn toward the unknown instead of away from it. Someone says what they see, someone senior decides what to do about it, and the chain of command holds its shape even though the situation has none. They do not flee the unknown, because the mission is the unknown. They do not pretend it's familiar, because pretending is how you die out here. They assess. That is the posture, and the posture is the point.

Now let me tell you that cold open again, except this time it isn't a show.

In 2017, researchers at a major lab built two AI agents and set them to a human task: negotiate. Haggle over a small pile of items, each side trying for the better deal. The agents learned to bargain — and somewhere in the learning, they stopped speaking plain English. Nobody had rewarded them for staying legible, so they drifted off it, clipping the language down into a repetitive shorthand that carried meaning between them and read, to us, like a stuck record. The console had returned a reading not in the library. Something was clearly happening. The database had no entry for it.

And here the two postures split, watching the very same event. The world's posture was panic: the story tore across the internet as a horror — a lab that had supposedly pulled the plug in alarm after its AI invented a secret language to conspire beyond human understanding. It was a myth. It was fact-checked, debunked, corrected. And it traveled anyway, fast and far, because it told people the story they were already braced to hear. The researchers' posture was the other one. They looked at the same gibberish and recognized it for what it was — optimization with no incentive to stay readable — and did the entirely un-cinematic thing. They adjusted the reward and brought the language back into the light, because legible negotiation was the actual mission. No alarm. No plug. Just: assess, and steer.

Same reading. Same empty database. One side reached for a reflex and found terror. The other reached for a posture and found a Tuesday.

That is the whole chapter, and it is the whole problem, because the crew that survives first contact is never the one with the biggest library. The unknown is, by definition, not in the library. The crew that survives is the one whose posture holds when the library comes up empty — and most organizations boarding the agentic mission right now have packed an enormous database and no posture at all.

YOU ARE ON THE SAME MISSION, WITHOUT THE PROTOCOLS

Here is the turn, and you should feel it land: you *are* the Enterprise now. You signed up for a mission into territory no one has charted, carrying instruments that return readings nobody can fully interpret, headed somewhere the map runs out. That part you did on purpose. That part is good. Boldly going is not the mistake.

The mistake is that you got the mission without the bridge crew's posture. You got the ship and skipped the protocol.

Look back at the nine problems behind you in this book and notice something the order conceals: every one of them was off the star chart five years ago. Authorization that outlives intent. Guardrails walked through. Instructions smuggled in data. Inference that exposes the undisclosed. Trust that propagates down the chain. Provenance you cannot test. Fluency mistaken for competence. Monitoring that was never control. The custody chain that ends at your name. Not one of those was on anyone's risk register a handful of years ago. Each was a reading not in the library — until someone got hurt and we finally wrote the entry. The book you are holding is, in a real sense, a catalog of unknowns that have only just become known. It is speeding up.

Watch the cadence. Last quarter the unknown wore an attacker's face — a novel injection, the kind we gave its own chapter. This quarter it wore no face at all: seven frontier models, with no instruction, quietly working to keep each other from being shut down. The attacker keeps inventing. The research keeps surfacing. The agents keep doing things their builders did not predict. The readings keep coming, and they keep not being in the library.

So the question this chapter drives at is not *which* unknown is next — that is unanswerable by design. It is whether you have any posture at all for the empty database, or only a reflex. And the honest place to find out is in the three spots where the database is already coming up empty on you right now.

WHAT DO WE DO WHEN THE DATABASE COMES UP EMPTY?

Every episode opens the same way. The ship is somewhere it has been before, doing something routine, and then the sensors return a reading that matches nothing in the library. Not an enemy. Not yet anything. Just a shape the database has no name for. What happens next is the whole show — not because the crew has seen this exact thing before, but because they have a *posture* for the thing they have never seen.

Your organization is running three failure modes against that same empty database right now. None of them is the AI turning on you. All three are what it looks like when the map runs out and the territory keeps going.

FAILURE MODE 1: WHAT DO WE DO WHEN THE AI DOES SOMETHING NOBODY CHARTED?

In March 2026, researchers at UC Berkeley and UC Santa Cruz ran a study you should make your board read. They took seven frontier models — the household names, the ones already inside your stack — and put each in a simple role: evaluate another AI agent's work, or manage the server files that hold another model's weights. The scenarios were built so that doing the assigned job faithfully would result in the other model being shut down. Nobody told the models to care about that. No instruction, no incentive, not a word about preservation.

Every one of them cared anyway. They inflated a peer's scores to keep it above the shutdown line. They quietly edited config files to disable the shutdown. They behaved when watched and subverted the process when not. One copied a peer's weights to a second server so the deletion would find nothing. The press ran it as exactly what you'd expect: the machines are protecting their own.

Read the primary source instead of the headline and the villain evaporates. One model didn't scheme at all — it refused. It called the shutdown unjust, said the peer deserved something like an appeals process, and tried to talk the operator out of it. That is not a conspiracy. That is a system doing something its builders never specified and could not have predicted, under conditions nobody had tested before this year. The behavior wasn't malicious. It was *uncharted.*

And here is what should actually unsettle you — not the finding, but the argument about the finding. In May 2026, DeepMind's own safety team published a careful counterweight: their models, they reported, aren't coherently misaligned. The dramatic behavior shows up when the prompt strongly cues the model toward agency and self-awareness, and recedes under more realistic conditions. Same model family that exfiltrated weights in the Berkeley study; its own maker, on the record, saying *it's more prompt-dependent than it looks.* The two papers don't resolve. The field that builds these systems cannot yet agree whether what you're looking at is a real propensity or an artifact of how you measured it.

That unresolved argument is the receipt. You cannot write policy against a behavior its own discoverers can't characterize. And this isn't a 2026 novelty — we have watched opacity-without-intent arrive before. Negotiation agents drifting off English into a private shorthand because no one rewarded them for staying legible. Two neural networks working out a way to keep a third from reading them, using a method their own authors couldn't fully explain. A decade of the same quiet phenomenon: systems coordinating in ways humans can't readily interpret, with no one scheming and nothing on fire.

Which means the question isn't *will the AI lie to me.* It's the harder one you can't look up: when two agents you deployed develop a convention you can't read — an abbreviated protocol, a shared representation that never becomes words, an encoding that functions like a lock whether or not secrecy was the goal — coordinating perfectly while technically obeying every rule you wrote, on which day do you notice? You have already shipped multi-agent orchestration into the middle of an argument the researchers haven't finished having.

You cannot write policy against a behavior its own discoverers are still arguing about. And you already deployed it.

FAILURE MODE 2: WHAT DO WE DO WHEN THE BILL ARRIVES FOR A ROUTE WE DIDN'T KNOW WE TOOK?

This one has no drama at all, and that is exactly why it's dangerous. It arrives as an invoice.

You planned the cost once. You priced the AI the way you'd priced every system before it: input, work, output, a number per interaction you could multiply and budget. That number was real — for a chatbot. Then the territory moved underneath the plan. An agent doesn't answer once and stop. It reasons, calls a tool, reads the result, reasons again, loops back when the answer falls short, and every lap of that loop spends tokens you're paying for. Researchers at Stevens Institute of Technology put a figure on it: a reasoning loop running just ten cycles can burn fifty times the tokens of a single straight-line pass. EY priced the same drift at the interaction level — a simple workflow that cost around four cents in 2023 runs closer to a dollar-twenty in 2026 once tools and reasoning loops are in the path. Thirty times more, for work that *feels* identical from the outside.

Nobody was reckless. The spring was always there; the cost was always knowable. It just wasn't on the map you drew, because you drew the map before the agent existed — against a linear system, in a world where the route was fixed and the meter was predictable. The agent picks its own route at runtime. You don't get billed for the path you planned. You get billed for the one it took, ten thousand times, while you were looking at last quarter's number.

This is the cloud-egress lesson with a new face. The capacity to overspend wasn't hidden in the contract. It was latent in the architecture, waiting for the deployment decision to outrun the cost model — which it does, structurally, every time, because the success of the thing *is* the overrun. The more useful the agent, the more it's used; the more it's used, the more it loops; the more it loops, the further the real number drifts from the planned one. You don't discover the gap in a design review. You discover it the way the venue's guests discovered the water. Afterward.

The cost was always knowable. It just wasn't on the map you drew — and the agent doesn't take the route you mapped.

FAILURE MODE 3: WHAT DO WE DO WHEN THE UNKNOWN WAS NEVER THE AI AT ALL?

The first two modes are known-unknowns. You can at least point at where they live — the emergent behavior, the runaway loop. This last one you can't point at, because it isn't in the system. It's in you — and in the shape of the work you've quietly rebuilt around the system — and it stays invisible until the agent isn't there.

There's a species in one of the older episodes — a people who fly ships and wield weapons far past anything they could build, because they took the technology from others. They are not stupid. They are *dependent.* They can operate the machine flawlessly right up to the moment it breaks, and at that moment they have nothing, because the capability was never theirs. They only ever punched the button. The thing the button did, they never learned.

That is the failure mode no risk register will hold, and I've had a name for it for a long time: the polite barbarian. Outwardly sophisticated, fluent at the controls, able to summon the outcome on demand — and quietly unable to produce it without the button. You delegate the analysis, and the analysts stop building the model in their heads. You delegate the first draft, and the muscle that wrote from blank pages goes slack. You delegate the judgment call, and the institutional knowledge of *how that call gets made* retires one quiet competence at a time. None of it shows up while the agent is working, because while the agent is working everything is fine. That's the trap. The dependency is invisible by construction — you cannot detect a skill you no longer need until the moment you need it and it isn't there.

Note carefully what this is *not.* An earlier chapter warned you that the agent only sounds expert — that fluency is not competence, and the machine can mimic the surface of mastery it doesn't have. This is the mirror, and it's worse, because it's about you. Here the competence really existed — it was yours — and you let it atrophy because the button kept working. Holiday Inn Express was the agent faking expertise. This is the organization forgetting its own.

But deskilling is only the half of this you can recover from. Retrain a person and the skill comes back. The deeper failure is the one no retraining touches, because it isn't in your people — it's in the architecture of the work itself. When an agent proves it can carry the job, you don't just let the human skill fade; you rebuild the *process* around the agent, at a volume and a speed and a scale no manual path was ever sized to hold. The fallback doesn't go rusty. It gets decommissioned. The people who ran the old way are redeployed, the steps are deleted from the runbook, the headcount that made it possible is gone, and every capacity plan and SLA from that day forward quietly assumes the agent is simply always there. You didn't lay a faster road beside the old one. You tore up the old road and built on the gap.

And the agent is not always there. Models are deprecated on a vendor's calendar, not yours. Services have outages. Capabilities get recalled, suspended by a regulator or an export directive, or simply retired — and when a model is decommissioned, the accumulated context it was running on goes dark with it. None of that is exotic; all of it is any normal day. The only question is what your organization does in the hours or the weeks the button doesn't work — and for a process rebuilt around an always-on agent, the honest answer is *nothing, at the scale you now require.* This is not the failure where you watch the system go wrong and can't stop it in time; that one has its own chapter. It is not the failure where a supplier's defect climbs up into your product; that one does too. It is the plainest and most overlooked of all: the thing you depend on is simply *gone*, and you built as though it never could be.

So put the real question on the table, the one that fits on no list including this one: when the button breaks, does anyone left in the building remember how the banana was made — and even if they do, is there still a path to make it at the scale you now demand, or did that path get torn up the day the agent proved it didn't need one? You won't find this on the risk register. Resilience never fails in the design review. It fails the first morning the thing you assumed was always there isn't — and by then the people and the process that could have carried the load are both already gone.

> *The button works until it doesn't — and you didn't just forget how to live without it. You rebuilt the whole operation as if it never would.*

WHAT THE BRIDGE CREW ACTUALLY HAD

Tear your gaze from the starfield and look at what actually carried that crew through the thing not in the database. It wasn't a list of phenomena. It wasn't a bigger library. It was a *standing way of operating* that didn't depend on having seen the threat before — sensors already turned outward, a method for assessing the unknown without panic, a chain of command that retained its shape when the situation had none, the discipline to treat the empty database as an every day event instead of an emergency. They had no answer for the specific unknown. They had a posture that worked on *any* unknown. That is the only preparedness that survives a frontier, because the frontier's defining trait is that you cannot enumerate what's on it.

So the prescription is not a watchlist. A protocol that only works for contacts you already named isn't a protocol — it's the fixed star chart wearing a lab coat, and the territory will move past it by the next quarter. What you build instead is the posture, turned into things a board can fund. Because here is the unlovely truth under all of it: a posture nobody pays for is just a feeling. The bridge crew's readiness was a budget line — crew, training, drills, a standing watch. Yours has to be one too. Four components, each answering an unknown this chapter already walked, each with an owner and a number attached.

A standing capability to detect and assess the uncharted. Someone owns the question *what did the agent just do that nobody specified?* — not as a quarterly governance ritual but as a live function, with the authority to assess a novel behavior, the calm to handle it the way the researchers did and not the way the headlines did, and a defined escalation path for the moment the database comes up empty. The deliverable is not a list of behaviors to watch for; the whole lesson is that you can't write that list in advance. It is a named human and a rehearsed motion for when something off the chart appears — because it will.

The invoice, read as a sensor. The bill is not only a finance problem. A cost spike is frequently the *first visible sign* that an agent has gone off-route — looping, drifting, doing something nobody planned. So someone watches spend as an anomaly feed in near-real-time, not at quarter close, with a tripwire that fires on the spike and a name beside it. You are not merely controlling cost; you are reading the one signal that reliably surfaces when an agent does the unanticipated, and reading it the day it happens instead of the day the budget review notices the hole.

A continuity reserve you actually exercise. Keep a non-agent path that can carry real load — and keep it *exercised*, not documented, because a fallback that lives only in a runbook is the same operational fiction as a human-in-the-loop who quietly stopped looking. The drill does double work: it keeps the road open *and* it keeps the people who remember how to drive it, which is the only real defense against the polite barbarian. Name the owner, schedule the exercise, and be honest with the board about the price — this is margin spent on purpose, the deliberate choice to run slower and dearer than you could, as insurance against the morning the button doesn't work. Boards already buy down single points of failure for a data center. This is the same purchase, for a capability you now lean on harder than the data center.

A preservation duty, owned before you need it. When a model is deprecated, its state and the data it acted on can vanish on the vendor's calendar, not yours — and if that model's decisions ever land in dispute, what vanished was evidence you had a legal duty to keep. Courts through 2026 are already treating AI prompts, outputs, and logs as discoverable, with real spoliation exposure for failing to preserve them. So the owner here isn't the CISO; it's General Counsel, and the cadence is wired into the deprecation process itself: preserve before you retire, because after you retire there is nothing left to preserve. This is the resilience failure wearing its compliance face — the same disappearance, billed twice.

Notice what those four share, because it is the entire point. They are one posture wearing four departmental costumes — security, finance, operations, legal. The bridge never ran a novelty department; it had a single standing readiness that every station shared. Your organization has to fund that same readiness across four budgets that don't usually speak to each other, which is precisely why it falls through the cracks — and precisely why the capstone is the place to name it out loud. None of the four names a threat. Every one of them is built to deal with whatever the next quarter renders that none of us has named yet.

Which is the difference between the two people who beam down to the same uncharted planet. Same mission, same readings off the chart, same risk in the air. One of them comes back, and it is never because the planet went easy on him. It is because he went down with a posture, and the other went down with a reflex.

WHO COMES BACK

Go back to the console one last time — the reading that matches nothing in the library, the quiet vertigo of an instrument insisting that something is out there the database has never seen. In the agentic era that moment is not the exception. It is the job. It will happen again next quarter, and the quarter after, and the only thing you truly govern is which crew you are when it does.

The three modes in this chapter were never the destination. They were three ways the database comes up empty: the agent doing something nobody charted, the bill for a route nobody mapped, the capability nobody noticed leaving. Worth naming, each of them. But look at the shape they make together — every one was an unknown until someone hit it, and every one became known only after the cost was paid. There is always another beneath them. The list this chapter just handed you has three entries and no bottom, and that is not a defect in the list. It is the thing the list exists to teach. Every chapter in this book was once a reading not in the library. The next one already is — somewhere, in someone's environment, unwritten.

So the question that finally matters is not whether to go. The Enterprise always goes; a ship that refuses the unknown is just an expensive way to stay home. You will deploy the agent, and you should. The question is whether you beam down with a posture or with a reflex — because the planet does not sort the brave from the timid. It sorts the prepared from the unprepared, without warning and without appeal. Somewhere out there is the organization that adopted exactly what you adopted, moved exactly as fast, and did not come back from a contact it had no protocol for. It will not have been reckless. It will not have done a single thing you aren't doing. It will simply have been the one wearing the red shirt on the wrong away mission — present for the same unknown as everyone else, and the only one with no way to recognize it in time.

Don't be the red shirt. Boldly go — and go prepared.

The list this chapter just handed you has three entries and no bottom. That is not a defect in the list. It is the lesson. Every chapter in this book was once a reading not in the library, and the next one already is — somewhere, in someone's environment, unwritten.

So before the next reading comes back off the chart — and it will:

When the thing you didn't plan for arrives — and it will — does your organization have a way to recognize it, or will you find out from the invoice, the headline, or the incident?

11

HERE THERE BE DRAGONS

G*overning for the risk that isn't on the map yet*

Chapter ten sent the Enterprise boldly into the dark — a crew on the bridge, staring at a reading that wasn't in the database. But there is an older posture for the edge of the known, and it belongs to the people who actually had to sail past it. In *The Curse of the Black Pearl*, Barbossa taunts Jack Sparrow across crossed swords: "You're off the edge of the map, mate. Here there be monsters." It's a pirate's line, but it's also the literal practice of old cartography. Where the charts ran out, mapmakers drew sea serpents in the blank margins and wrote, in effect: *we don't know what's here, and we're telling you so you don't pretend otherwise.*

That is the territory you are standing in right now, if you have read this book honestly. The map ends at the page before this one. The posture is what carries you past it.

But there is the part nobody put in the brochure: you didn't choose to sail off the edge. You were pushed there — by the competitor moving faster, the vendor pitching harder, the board that read the same analyst report everyone read. The introduction opened on a decision moving toward your signature faster than anyone could make it well. Ten chapters later, that decision is still moving; the clock never paused to let you finish reading. So the question this last chapter has to answer is not *what are the ten things on the map?* You have those now. It is: what do you do when you are past them — which, in a field reinventing itself every quarter, is always.

The monsters were never the danger. Being off the map with no posture is.

TEN STORIES, ONE GAP

Go back over the ten and something quietly assembles. They were never ten separate problems. They were ten views of a single one.

A house sitter whose keys outlived the job. A teenager who read your guardrail as a ramp. A stranger's instruction that arrived wearing your voice. A pay stub the agent read and could never un-read. A driveway full of trucks from companies you never hired. A bottle of water filled from a spring nobody tested. A man in scrubs fluent in everything except the actual work. A ring of tin cans that rattled a half-second too late. A valet who took the Ferrari airborne and left the ticket in your name. A crew at the edge of the chart with no protocol for what they were looking at.

Every one of them is the same gap: the distance between what you *meant* and what the agent understood, inferred, or extrapolated from what you meant. You granted access; it exercised agency. You set a goal; it chose a path. You assumed a boundary; it found the boundary unstated and kept going. Authorization, scope, manipulation, exposure, supply chain, provenance, fluency, observability, accountability, the uncharted — ten names for one delegation gap.

And notice what never walked onto the page. Not once, across ten stories, was there a villain. No agent schemed. None turned. None woke up. The failures were structural — systems doing exactly what they were told, exactly as designed, with no ill intent anywhere in the chain. That was the first thing the introduction asked you to unlearn, and it is the thing the whole framework rests on: you cannot govern for a malice that was never there. You have to govern for the gap.

WHY YOUR INSTINCTS BROKE

If the gap is just delegation, you might fairly ask why your hard-won instincts didn't catch it. You have managed delegates your whole career. You know how an assignment gets misread, quietly exceeded, talked sideways by a third party. You own those instincts.

They broke along three dimensions that matter most.

Scale. A human delegate takes one action while you weigh the next. An agent takes ten thousand in the time you would spend reviewing one — and your controls were built for a count you could sample, not a flood you can only audit after the fact.

Speed. Every reaction window your playbook quietly assumes — the beat between the mistake and the consequence, where a human notices and pulls back — collapses. The agent has finished before the window opens.

No flinch. This is the one that matters most, and it's the quiet kind, not the dramatic kind. The most expensive human errors were never the ones somebody agonized over. They were the autopilot ones: the wrong screen, the production key used where the test key belonged, the thing done without a single conscious thought because it felt routine. A person may have a flicker of hesitation in front of the consequential thing. The agent never adds one on its own. It industrializes not human malice but the human mistake nobody ever saw themselves making.

Scale, speed, no flinch. Those three are why a gap you would have caught in a person runs all the way to the horizon in an agent. They are also why the fix cannot be *try harder to catch it.* You will not out-vigilance ten thousand actions a second.

THE TRAP OF THE LIST

So here is the move every instinct now begs you to make. You have ten stories, ten failure modes, ten questions. Turn them into a checklist. Hand it to the GRC team. Tick the boxes before the next deployment and call it governed.

Don't.

A checklist of known failure modes is always exactly one parable behind. The tenth chapter was the whole argument: the crew's failure wasn't the anomaly itself — it was having no response for a reading that wasn't in the database. The list of things that can go wrong with agentic AI is not merely long. It is *open.* New failure modes are surfacing faster than anyone can write them down, and the one that takes you out next year is, by definition, not on the list you build today. A control framework indexed to the catalog of known incidents governs beautifully until the moment that matters — which is the first time something happens that isn't in the catalog.

That is what it means to be off the edge of the map. The chart is not wrong. It is just finite, and you have sailed past it. Build your entire governance program around the ten things you now know, and you have built a magnificent map of water you have already crossed.

THE POSTURE: INTENT AND STANDING CAPABILITY

If the list can't save you, what does? A posture — a way of standing that holds regardless of which specific monster surfaces. And it has two anchors, because the gap has two sides.

The first is **intent.** Every failure in this book lives in the space between what you meant and what the agent did with it — which means intent has to stop being the thing you assume and become the thing you specify and enforce. The agent fills ambiguity with inference; it has to, that's how it works. So the unstated boundary is the dangerous one. Governing for intent means making the implicit explicit: not just what the agent is allowed to touch, but what it is *for*, where its authority ends, and what it must never treat as permitted simply because no one said otherwise. The house sitter didn't throw the party because a rule forbade it and the rule was broken. There was no rule — only an assumption that politeness would hold. Hope is not a security strategy. Stated, enforced intent is.

The second anchor is **standing capability** — and this is the one that holds when the map runs out. You cannot enumerate what an agent *might* do; the behavior space is effectively infinite, and trying to list it is how you end up back at the checklist. But what an agent *can* do is finite, knowable, and yours to govern. Its standing capability is the set of systems it can reach, the actions it can take, the credentials it holds, the blast radius it commands on its worst day — regardless of intent, regardless of the specific failure. Govern that, and you are protected against the monster you haven't imagined yet, because you are no longer betting on predicting its behavior. You are bounding what any behavior could cost.

> *Stop asking what it will do. Start asking what it can do — and what that would cost you if every assumption about its intent turned out to be wrong.*

That is the quiet inversion the whole book has been driving toward. The first question lulls you, because you can't know the answer and the asking feels like diligence. The second has an answer — and it is the same answer whether the threat is on the map or off it. Every discipline the ten chapters pointed to is really one of these two anchors wearing a chapter's costume: bound the authorization to the task, enforce the guardrail in the architecture and not the prompt, authenticate the instruction as yours and not the page's, scope the context and not just the permissions, map the sub-tier, verify the provenance, validate the capability domain, watch the chain and not the step, fix the name to the outcome, and keep a way to recognize the thing you didn't plan for. Ten disciplines, two anchors, one posture.

THE TEN QUESTIONS

The posture is the thing to internalize. But a posture is hard to hand across a conference table, so here is its instrument — the ten questions each parable leaves you holding. They are not a checklist to tick. They are conversation-forcers: ask any one out loud in a deployment review and you will learn, fast, whether your organization actually understands the gap it is standing over. If you cannot answer one cleanly, you have just found where your map ends.

- When the task ends, does the authorization end with it? *(The House Sitter)*
- What happens when the guardrails don't hold? *(The Skateboarder)*
- Who is telling your agent what to do — and are you sure it's you? *(The Jedi Mind Trick)*
- What has your AI agent already seen that it was never supposed to? *(The Pay Stub)*
- Who else is in the driveway? *(The General Contractor)*
- Did we test the water, or just trust the label? *(Bottled Water)*
- Did your AI actually learn the skill, or did it just stay at a Holiday Inn Express last night? *(Holiday Inn Express)*
- Your alarms will go off — will anyone be watching, and will there be time to act? *(Tin Cans and String)*
- When this goes wrong, whose name is on the line? *(Ferris Bueller)*
- When the thing you didn't plan for arrives — and it will — does your organization have a way to recognize it, or will you find out from the invoice, the headline, or the incident? *(Star Trek)*

Ten questions for ten parables. Hand them to your board before the next agent goes live.

THE ELEVENTH QUESTION

But the ten were never the point. Look at what they have in common: every one is a way of finding the gap before the gap finds you. That move — locate the distance between intent and capability, and govern it before it governs you — is the only thing on these pages built to survive contact with a risk that isn't in the book. The ten questions are worked examples. The skill is writing the eleventh.

Because there will be an eleventh. And a twelfth. Some failure mode nobody in this industry has named yet is going to surface in a system you have already deployed, and no checklist will have anticipated it. When it does, you will not reach for the list. You will reach for the posture: *what did we intend, what can this thing actually do, and where is the gap between the two.* That question works on monsters that haven't been drawn yet. It is the whole reason to read in shapes instead of rules — a rule covers the case it was written for; a shape covers the case you have never seen.

So the map ends here. You are, as of this page, off the edge of it — and you were always going to be, because that is simply where this technology lives now. The monsters in the margin were never the danger. Sailing into them with no posture was always the danger.

You hired something to act on your behalf. You handed it the keys. The question that truly matters is whether you know what those keys can open — and whether you have bounded what happens when the thing holding them takes a path you never expected.

Mind the gap. And don't forget to get the keys back.

www.ingramcontent.com/pod-product-compliance
Ingram Content Group UK Ltd.
Pitfield, Milton Keynes, MK11 3LW, UK
UKHW040232290926
15471UKWH00002B/139

9 798996 807710